DIGITAL DIRECTORS NETWORK

The DOMINO Guide:

The Definitive Boardroom Guide On Digital, Cybersecurity And Systemic Risk Governance

Bob Zukis

CEO, Digital Directors Network

2025

The DOMINO Guide: The Definitive Boardroom Guide on Digital, Cybersecurity and Systemic Risk Governance:

www.digitaldirectors.network

ISBN: 978-1-7350430-3-6 (Paperback)

Suggested reference: Zukis, Bob. "The DOMINO Guide: The Definitive Boardroom Guide on Digital, Cybersecurity and Systemic Risk Governance," DDN Press, Manhattan Beach, 2024

About Digital Directors Network

Digital Directors Network (DDN) is the premier boardroom network of IT and cybersecurity executives, corporate directors and organizations working together to shape and secure the digital future. Together we are advancing the practice and profession of digital, cybersecurity and systemic risk governance. DDN was founded in 2017 and its members represent many of the world's leading corporate directors and executives working on both sides of the boardroom table to solve these challenges.

About DOMINO And This Guide

This edition of The DOMINO Guide reflects the knowledge created and shared at DOMINO 24, the leading boardroom executive learning event focused on advancing the practice and profession of digital, cybersecurity and systemic risk governance. The DOMINO Guide is a reference and implementation guide that reflects the steps and leading practices that any board can take to strengthen their role in how the complex digital business system creates and sustains business value.

DOMINO 24 was held at The University of Chicago Booth School of Business in Chicago, IL. This executive learning experience convened policy leaders, corporate directors, digital and cybersecurity executive leaders and experts to network, teach and learn about solutions that strengthen the role of the board in shaping and securing the digital future.

A combination of business school caliber classroom-based lectures, peer learning exchanges, and deep interactive masterclasses led by subject-matter experts, corporate directors, and industry leaders, DOMINO 24 was focused on advancing solutions that develop digital and cybersecurity governance as a high-performing part of boardrooms around the world to strengthen the board's role in the digital business system.

DOMINO 24 provided practical, and actionable insight and instruction to continue to

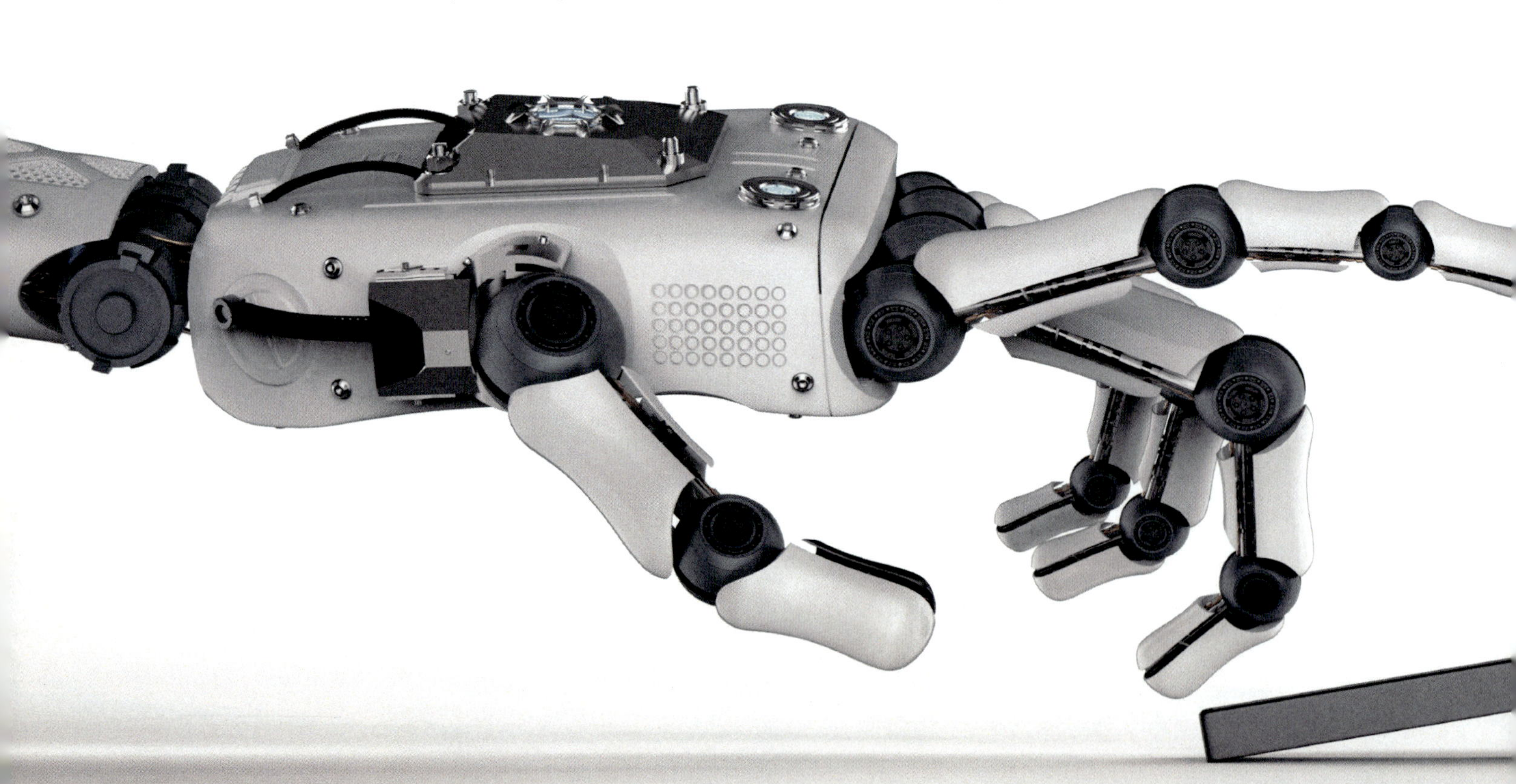

build upon the pioneering work of Digital Directors Network and its membership.

Executive leadership from SolarWinds provided the opening keynote at DOMINO 24 to set the tone and share what they have learned from a major cybersecurity incident that woke the world up to the threat of systemic cyber risk along with the emerging forces of government scrutiny and accountability.

The U.S. Cybersecurity and Infrastructure Security Agency (CISA) Executive Assistant Director for Infrastructure Security Dr. David Mussington presented and answered questions on the work being done at the senior most levels of the U.S. government to secure complex digital systems against the growing unpredictability of worldwide threats.

Other policy leaders, executives and directors from Target, Delta Airlines, PNC Bank, Barnes Group, WD-40, Kraft Heinz, EY, Dominion Energy, Campbell's, United Airlines, and many other leading companies enriched the DOMINO 24 executive learning experience with their perspectives and insights.

Enriching the overall learning experience were DOMINO 24's learning partners who shared their deep insights through a masterclass series and peer learning exchanges focused on solutions to the challenges of digital and cybersecurity oversight. They included White & Case, EY, Veracode, Proofpoint, X-Analytics, Kudelski Security, Corporate Board Member, CYVERSITY, Equilar, and Hitch Partners.

Four conference attendees were recognized for their contributions to the practice based upon their thought leadership submissions to the *DOMINO 24 Call For Papers*. Submitted papers were peer reviewed and four were selected for their contributions to the body of practical knowledge on digital, cybersecurity and systemic risk oversight. Their papers are included in the Appendix.

Pre-event participants also attended the widely acclaimed *DDN QTE 501 Boardroom Readiness Masterclass for IT and Cybersecurity Executives*. All attendees enjoyed opportunities to network including our Veteran's and Women's leadership breakfasts, an evening reception, dinner and an entertaining display of complex systems of a different kind.

DOMINO 24 was CPD Certified as an executive learning event and attendees earned 13 hours of CPD credit. Participants who attended the *QTE 501 Boardroom Masterclass for IT and Cybersecurity Executives* on May 14, 2024 earned another 13 hours of CPD.

CPD CERTIFIED
The CPD Certification Service

DOMINO: Solving The Digital, Cybersecurity And Systemic Risk Governance Puzzle

The digital world is all around us. The digital future will be what we make it — or what we allow it to become if it remains ungoverned.

The corporate boardroom has a vital role to play in shaping and securing the digital future and our mission at DDN is to make sure that every corporate board is a high-performing part of their company's digital journey. Together with our members, this is the problem that we are solving — and we know that solving it drives significant business value — and we know how to solve it.

But the digital future is stumbling forward as it is being tripped up by old, new and different types of risks. Boardroom leadership and capability on these issues is often lacking which creates weakness throughout the system. This weakness impairs the economic growth, output and resiliency of businesses and economies around the world.

DOMINO 24 was focused on accelerating and scaling the solutions that are helping boardroom and executive leaders strengthen directorship and the role of the board in shaping and securing the digital future.

While boardroom awareness of digital innovation, cybersecurity and systemic risk is as high as it ever has been, the strength of the boardroom as a control is far weaker than it needs to be.

The advent of artificial intelligence (AI), high profile cybersecurity incidents and new regulations from the U.S. Securities and Exchange Commission and in the EU have delivered many different wake-up calls. But not enough is being done, fast enough, to strengthen the critical capabilities needed from boardroom leadership. Remarkably, there are also some organizations working against common sense board reform on these issues, making our job collectively harder but the job of the hackers easier.

The cybersecurity failings of the corporate boardroom were recently on full display during the largest cybersecurity incident in American healthcare history at UnitedHealth Group (NYSE: UNH). During Congressional testimony from the UNH CEO. U.S. Senator Ron Wyden attributed this cybersecurity failure to a lack of boardroom leadership pointing to the fact that there is no director cybersecurity expertise on the UNH board. Directors with cyber expertise addresses one key part of solving this puzzle, but much more is needed.

DOMINO's world-class learning experience helps corporate directors and executives rise to this digital leadership moment to learn how to lead their companies safely and securely into the digital future. Focused on governing digital value creation and protection, the good news is there are many corporate directors, digital and cybersecurity executives already leading in the boardroom on these issues. **And evidence shows that when corporate directors lead on these issues, positive business impacts follow.**

In a highly connected world, risks are distributed

across complex and interdependent digital ecosystems and companies. But the lack of a collective security model or leadership mindset means that one company's risk or failure, can often be a problem for many companies, e.g., CrowdStrike.

And too much is at stake for the corporate boardroom to not be leading on these issues. At DDN and DOMINO we are creating a collective understanding and mindset in the boardroom and building the capabilities so that every boardroom can fulfill its leadership role and obligations in the digital world.

Thank you to the DDN Members who attended DOMINO 24, our speakers, content partners, our boardroom certified Qualified Technology Experts (QTEs) and the other leaders who are stepping up to this challenge and leadership moment. The digital future needs you and will be in better hands because of you.

A special thank you to the DDN Advisory Board comprised of Andrew Chrostowski, Tony Cole, Fay Feeney and Jerry Nowicki for their advice, counsel and support.

***The DOMINO Guide* is produced from the knowledge created and shared at DOMINO 24, the leading boardroom and executive learning event on digital and cybersecurity governance. It is both a reference guide and implementation guide and we encourage you to share it with your leadership team and entire board.**

As the definitive voice on these issues, DDN's members and our DOMINO 24 attendees are collectively building the practice and profession of digital, cybersecurity and systemic risk governance — making directorship a high-performing part of the system leading companies successfully and safely into the digital future.

Thank you for being with us on this journey.

Bob Zukis
DDN Founder and CEO
bob@digitaldirectors.network

Executive Summary

Evidence shows that effective digital, cybersecurity and systemic risk governance creates and protects business value. Empirical evidence proves that significant positive business impacts and tangible business results are created when there is a high-performing corporate board in place with the expertise to capably govern the digital upside and protect against the downside. Negative impacts follow when this is absent.

Voluntary and mandatory standards are developing. Voluntary and mandated standards from regulators that are specific to the role of the board in governing digital business systems are maturing and emerging. Regulatory coercion is forcing the adoption of specific boardroom policies, processes and procedures that are strengthening the role of directorship in the digital business system. Leading practices standards continue to develop and mature.

Self-regulated board transformation remains the best path forward. The leading edge of digital and cybersecurity governance is now being self-regulated into place. The number of boards and corporate directors who are transforming one or more aspects of their governance systems continues to grow. While self-regulation is slower than forced government mandate, these boardroom leaders recognize their responsibility to shareholders and stakeholders and are taking action. Regulatory mandate also tends to lag the reality of market risks.

Digital risk is rapidly changing, expanding, and is not sufficiently understood. New risks continue to emerge as a result of new technologies and the growing complexities of the complex digital business systems that power companies. Corporate directors are learning about these weaknesses the hard way — because of incidents at their company or high-profile companies like UnitedHealth Group and CrowdStrike. Technologies like AI are creating new risks, and the understanding, identification and mitigation of systemic cyber risks like the CrowdStrike incident is nascent and not keeping up. More large scale incidents are guaranteed.

Solutions exist, they are just not widely understood or distributed. We know how to fix the problem as it is well understood by DDN and the leaders who are at the forefront of implementing processes that have been proven to work to drive and protect business value. However, more leaders on both sides of the boardroom table need to be proactive and willing agents of change and the pace needs to accelerate. Individual leadership initiatives are slow and new stakeholders need to step-up including institutional investors and the corporate leaders in IT and cybersecurity who have an ethical and moral responsibility to fix the problems that their innovations have created, starting with strengthening boardroom leadership over these technologies. created, but a business opportunity if they do.

The Value Of Digital, Cybersecurity And Systemic Risk Governance

Albert Einstein has famously said that if he had an hour to solve a problem, he'd spend 55 minutes thinking about the problem and 5 minutes thinking about the solution. The truth of his statement reflects the reality that in order to solve a problem, you first have to understand it.

Over the last eight years, DDN has worked exclusively to understand the role of the boardroom in shaping and securing the digital future and the importance of strengthening directorship as a control in optimizing the potential of the digital future. Together with our corporate, executive and boardroom members we know how to solve this problem and what's at stake when we do, and don't.

Based upon a growing body of evidence, we know that strengthening the boardroom works to create and protect the vast amounts of business value derived from complex digital business systems. That value drives economic output and growth, i.e., revenue and profitability and many different corporate value propositions.

And strengthening corporate governance strengthens the whole digital business system to create a resilient system that delivers and protects investor and stakeholder interests.

The debate is also over on these benefits, but denial is not. Boardroom leadership drives significant positive business impacts, the evidence supports this. The lack of boardroom leadership does the opposite and the proof makes almost daily headlines.

Our mission at DDN and DOMINO is to scale, distribute and continue to advance these solutions. By strengthening the boardroom as a control in the complex digital business system we enable boardroom leaders and management teams to stop being literal hostages to their digital business systems. Whether being held captive by ransomware, a malicious attack, or even human error, our mission is to set them free so that they can focus on navigating their companies safely and securely into the digital future.

Empirical evidence and research from MIT shows that effective digital and cybersecurity governance creates significant positive business impacts. These benefits include higher revenue growth, profitability, return on assets, and market capitalization. The business impacts their research identified when boards have a critical mass of true digital capabilities and expertise included:

- 38% higher revenue growth.
- 34% higher market capitalization growth.
- 34% higher return on assets.
- 17% higher profit margins.

Their conclusion is that boardroom leadership delivered by corporate directors with digital expertise drives material business results.

Research from Virginia Tech during the SEC's comment period for their proposed cybersecurity disclosure rules also gathered evidence of the positive impact to cybersecurity risk when there is director cybersecurity expertise on the board, and the negative impacts when it is lacking. These impacts included:

- Lack of director cybersecurity expertise leads to superficial check-the-box oversight.
- Boards without cybersecurity expertise rely too heavily on the CISO which can create a circular oversight environment that lacks independence or even encourages CISO's to water-down problematic issues.
- Director cybersecurity expertise can strengthen the effectiveness of the CISO with the full board and the C-suite.

- The presence of director cybersecurity expertise enables directors to provide proactive, value-added oversight of cyber-security risk that would not be possible without it.

The conclusion is that Boardroom leadership delivered by corporate directors with cybersecurity expertise creates and protects business value by reducing cybersecurity and systemic risk.

Whether creating business value or protecting it, the evidence shows the positive business impacts that effective digital and cybersecurity governance provides while reinforcing the board's central role as a control in the digital business system. From digital value creation to cybersecurity risk mitigation, this research along with DDN's observations and the experiences of our DDN members also support these findings.

Boardroom leadership matters. **Strengthening the boardroom as a leadership control in the complex digital business system strengthens all controls — that is its superpower.** Strengthening the boardroom strengthens the entire control system. It is the most powerful leverage point available to drive digital innovation and cybersecurity resilience. It is also a superpower that is readily available to every board and a high ROI solution that can be efficiently and quickly implemented.

At DDN, together with our members, we know how to solve this puzzle. Read on to learn how you can too.

The Systemic Power Behind DOMINO's Puzzle

DOMINO 24 convened business and boardroom leaders who are at the forefront of creating and implementing the processes that are key to solving the challenges facing boardrooms in digital, cybersecurity and systemic risk governance.

These leaders are solving these issues by focusing on three areas:

1. **Standards:** They are leveraging and applying existing digital and cybersecurity governance standards and creating new ones.
2. **The System of Governance:** They are implementing adaptive systems of corporate governance with the proper director skills and expertise, boardroom structure, and a comprehensive scope of risk oversight.
3. **The Scope of Digital Risk Oversight:** They are focused on systemic resiliency by understanding and monitoring opportunity risk, cybersecurity and systemic risk throughout the digital business system.

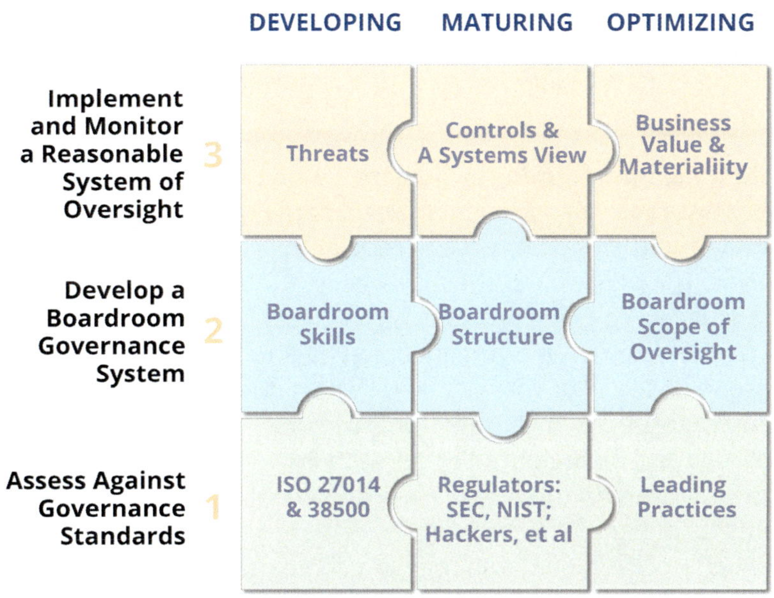

With so much economic output and growth now dependent upon digital business systems, solving this problem is a core responsibility for every corporate boardroom whether public or private, U.S. based or not. Governing digital, cybersecurity and systemic risk cannot be delegated to external third parties. The board needs the director and boardroom capabilities to lead on these issues. While outside experts have an important and expanding role to play on these issues at the direction of the board, fiduciary duty cannot be outsourced.

> Warren Buffet has been quoted as saying that "risk comes from not knowing what you are doing."

This is also a lesson that was previously learned the hard way during the financial reporting crisis of 2002 when an over-reliance on third parties in accounting and financial reporting created a financial reporting integrity crisis that threatened the capital markets.

Regulated boardroom reforms were mandated with the Sarbanes-Oxley regulations which forced corporate boards to change their structure and take responsibility by having the levels of expertise needed on the board to govern these issues effectively — and it worked.

This three-layered framework, referred to as DOMINO's Puzzle, is the blueprint for strengthening the board as a control on these issues. As a structured way to understand the systems needed to oversee these issues, it is the only purpose built model focused on helping any board conceptualize what they need to do to strengthen their effectiveness in governing complex digital business systems.

The three layers and nine parts work together systemically to enable any board to:

- **Leverage and implement established standards:** Others have taken big steps towards solving these problems and are continuing to advance effective solutions — they can be learned from. Recognizing that there is a strong body of existing work and guidance that has been developed by peer leaders that is proven to work is Step 1. Government regulators are not on the leading edge of these issues, it is boardroom and executive leadership around the world who are solving these problems. The DOMINO 24 Guide reflects key updates on some of the most useful standards.

- **Build a high-performing and adaptive system of governance:** Digital technologies emerge quickly, and their risk landscape constantly evolves. The only way to effectively govern this kind of dynamic environment is by building an oversight system that is designed and built to adapt to this level of volatility. This system needs to be comprised of digital and cybersecurity expertise and be organized effectively within the larger structure of the board. Finally, the board needs to implement a comprehensive system of oversight focused on risk in the right contexts, and then effectively monitor those risks.

- **Build systemic resiliency by understanding the different types of risk created by the complex digital business system.** Digital risk is not a generic risk. It is unique due to the nature of digital business systems including their many different inter-connected technologies and parts, and the specific risks that surround them. It requires new skills, approaches and understanding in the boardroom and a focus on the three different aspects of risk created by the complex digital business system: opportunity risk, cybersecurity risk, and systemic risk.

DOMINO's Puzzle is a market derived framework that reflects what is already happening and what is working to strengthen the boardroom as a digital control. DOMINO's Puzzle is also adaptive to developments in leading practices and regulatory standards as these issues mature. It applies a systemic view to create a sustainable model that will survive the test of time and future technology innovations.

Systemic resiliency is needed by the boardroom and within the digital business system to successfully deliver on the promise of the digital future.

Leading Practice: Recognize that this is a systemic problem, that needs a systemic solution. Solve the problem by breaking the plan for strengthening the boardroom as a control in the digital, cybersecurity, and systemic risk systems into its component parts of *standards*, the *board's governance system*, and the *system of oversight*. Use DOMINO's Puzzle as your framework and to benchmark your relative maturity against this market derived leading practices model.

1. The Standards And A Regulatory First, New Frameworks And Leading Practices

A significant amount of work has already been done to solve the challenges of governing complex digital business systems by boardroom leaders around the world. These solutions are manifested through the standards that already exist and that are continuing to develop. These standards reflect the leading practices, policies and processes that are already proven to work.

Frameworks are structured ways to analyze an issue or understand a problem. They are developed from a deep understanding and experience with an issue. The practice and profession of digital and cybersecurity governance is no longer a nascent part of corporate governance — it is now a developing part of high-performing boardrooms supported by a growing body of evidence that demonstrates their effectiveness in driving and protecting business value.

However, these practices are not yet reflected in all, or even most, corporate boardrooms. And there are also those working against common sense reforms on these issues. As these standards continue to develop and their implementation becomes more widely understood and distributed, the empirical evidence that proves their effectiveness will also continue to grow. This will ensure that they become a critical part of corporate governance policy and practice around the world.

Leading Practices From Cybersecurity Regulation

The opening keynote at DOMINO 24 was delivered by SolarWinds (NYSE: SWI) executive Brian Goldfarb. During DOMINO 24 most of the SolarWinds executive team was presenting oral arguments in court related to the groundbreaking lawsuit filed against SolarWinds in 2023 by the U.S. Securities and Exchange Commission related to their 2020 cybersecurity incident.[ii]

In addition to now settled shareholder derivative and class-action lawsuits, the SEC sued SolarWinds and their CISO for fraud and internal control failures. This litigation drew significant attention given the unprecedented inclusion of the SolarWinds CISO in the lawsuit. Several key issues in the lawsuit were dismissed in July 2024. These legal precedents will create new standards of practice and policy through regulatory interpretation and enforcement.

In addition to SolarWinds creating legal precedent for accountability in cybersecurity, the SEC's lawsuit follows the release of new cybersecurity disclosure rules that were finalized by the SEC in July 2023. These rules went into effect for most companies in December 2023.

Government regulators are increasing their oversight of cybersecurity both at the management and board level by adding new regulatory requirements. This added transparency and accountability raises the importance of the role of the board, directors and specific management actions on these issues.

Regulatory coercion through new compliance requirements is a very effective driver of short term and widespread change. However, regulation frequently lags the reality of market risks. It often only comes long after the need for remediation due to the negative impacts that have been realized in practice. Regulatory bodies around the world have done little to nothing to force corporate governance reform on digital, cybersecurity or systemic risk governance. That is starting to change.

Whether a board transforms their policies and processes through self-regulation, in reaction to a hack, or through government regulation and coercion, a tipping point is getting closer on these issues because of these converging forces — and it is long overdue.

By sharing their experiences with the DOMINO 24 attendees, SolarWinds leadership is helping others learn as their processes, policies and procedures for cybersecurity risk management and governance develop and mature. This type of leadership is commendable as we work towards building a collective security model and leadership mindset in the boardroom.

Brian was interviewed by JC Vega, Director at Sorenson Communication and a DDN Boardroom Certified Qualified Technology Expert (DDN.QTE). Brian started his presentation with an update of the SEC litigation including one important self-regulatory action that they have taken that has had a transformative positive impact between the board and management. **Almost immediately after the incident, the new SolarWinds CEO added a Technology and Cybersecurity Committee to their board.**

Prior to this, cybersecurity oversight was the responsibility of the Audit Committee for a period of time and then the Nominating and Corporate Governance Committee assumed responsibility. Court documents indicate that the full board had not received a presentation on cybersecurity for a two-year period preceding the incident.

While this boardroom organizational change came after the fact of the incident, that is not uncommon. Board reform often follows a major cybersecurity incident, making hackers a powerful, expensive and de-facto force as boardroom regulators. Although any board can materially improve their corporate governance in the short term by adopting leading practices in digital, cybersecurity and systemic risk oversight through self-regulatory action.

HEARD AT DOMINO

Investors and CEO's should be insisting that boards strengthen their role as a cybersecurity control — particularly CEO's. Why would any CEO want to go it alone with board that does not have a director with cybersecurity expertise covering their back?

The SolarWinds cybersecurity incident of 2020 was yet another of many wake-up calls that directors and executives have been slow to react to. This incident introduced many leaders to a frightening, but long present, dimension of cybersecurity risk — systemic risk. And one very vulnerable aspect of this type of risk — extended supply chain vulnerabilities. The SolarWinds hack by design, targeted a point of leverage within the SolarWinds digital system that deployed the hack efficiently into their customer base.

The CrowdStrike cybersecurity incident of 2024, while not the result of a malicious attack, is also an example of the inherent levels of systemic risk that exists in complex digital systems that can be triggered by simple error or systemic exposure within the system.

Systemic risk is the risk that exists in and between the parts of a complex system that can threaten the purpose of the system. Hacking SolarWinds software update process to gain efficient access and scale deployment leverage throughout their customer base is a perfect example of how hackers turn the system against itself to create widespread systemic damage.

> Defining and teaching how to identify systemic risk within the complex digital business system has been a part of the DDN mission since its inception. It is a core part of the DDN frameworks, our executive education programs and board-room certifications. We have trained and certified over 600 executives and corporate directors from leading companies around the world on these frameworks.

Systemic risk is a rapidly expanding threat in cybersecurity because of the inherent systemic complexities within digital systems that make connected business ecosystems vulnerable to this fast moving problem. Cybersecurity and systemic risk are unique risks that require new types of director expertise and new boardroom approaches to govern them effectively.

What lessons can be learned from the first-hand experiences of the business and boardroom leaders living through these issues? Key insights shared from SolarWind's Brian Goldfarb include the following:

- **Plans are worthless. Planning is priceless.** Plans are scripts outlining how an event or incident should unfold. However, incidents never go according to plan as threat actors are active participants who influence the outcome. The system often comes into view in new ways as an incident unfolds, uncovering unexpected aspects of the incident and its impacts. If an event deviates from the plan, the plan can quickly become ineffective unless people adapt swiftly. The planning process involves considering multiple courses of action and scenarios to prepare an organization for the unexpected. The act of planning builds muscle memory, experience and adaptability that enables an organization to respond in a controlled manner, to an uncontrollable situation.

- **There will always be unknown, unknowns.** There are things we are neither aware of nor understand in cybersecurity. That is one of the things that makes cybersecurity risk unique from many other types of risks. This means we cannot apply old models to governing or managing it. There is a new focus on third-party risk within the supply chain because of the SolarWinds incident. Potential gaps or blind spots in an organization's risk registry and mitigation plans always exist because the digital tools constantly change, and they bring new risks with far ranging systemic implications that are often not recognized until they are realized.

- **Speed of trust has to be built up front.** How cybersecurity incidents proliferate and propagate, and their instantaneous impacts requires supply chain and ecosystem trust to be built beforehand. This was vital for SolarWinds in retaining customers. Incident response requires the involvement of many internal and external stakeholders, so you need to plan for this. While malicious code spreads at machine speed, human response and resiliency operates with the *speed of trust*. This is one of the best business investments any company can make with their stakeholders in cybersecurity.

- **From the boardroom down, roles and responsibilities are changing.** This is a new ballgame, and old approaches are not good enough. The exchanges, expectations and relationships between the CISO and directors must evolve, and quickly — and on both sides of the boardroom table. The Technology and Cybersecurity Committee added to the SolarWinds board post-incident was a game changer in driving this change. New problems, which is what we have with these technologies, need new solutions. And that includes how the board gets its job done — they are a vital and critical part of the team and the system.

Experience is always the best teacher — but it can be a costly lesson. SolarWinds, UnitedHealth Group, MGM, Clorox, Caesars, Microsoft, CrowdStrike and many other companies are the examples of what can and will go wrong.

Leading Practice: Prioritize self-regulation as the most effective approach to strengthening the role of the board in digital, cybersecurity and systemic risk governance. By identifying the boardroom as a key control in the complex digital business system, both with regard to how the digital system creates value and how to protect that value, a strong case can be made for board reform. Ultimately, strengthening the boardroom as a leadership control over the digital business system, strengthens the entire system and evidence shows that this drives business value.

CISA And The Need For A Strong Public-Private System

On Day 2 of DOMINO 24, Dr. David Mussington, the Executive Assistant Director for Infrastructure Security (ISD) at DHS CISA, gave an enlightening and eye-opening speech about the realities of the cyber threats we are facing as a country.

Dr. Mussington talked about the threats facing our critical infrastructure which can impact every company and all of us. He emphasized not just the potential opportunities for further public and private collaboration, but the critical need for it. During his speech, Dr. Mussington spent some time talking about the threat actor *Volt Typhoon* and the risk to our critical infrastructure. He stated that this threat is living within our critical infrastructure and is "living off the land" by utilizing our own IT tools against us by using them to gain and continue long term unfettered and mostly undetected access — a systemic risk.

This tactic is a systemic threat that leverages the systemic weaknesses within the system to escalate and distribute the threat throughout the system. He also brought up the adversarial tactic of dismantling response capabilities far in advance of an attack. A tactic that impairs the ability to recover and respond in a timely manner. This type of strategic and persistent tactic is another reason cybersecurity is not like other enterprise risks.

He discussed how our water, power, sewage, and many other systems are already compromised and that companies need to work with CISA to understand how the adversary is doing this and how to expel them from our networks.

The active, well-funded, and intelligent adversaries behind these threats are also other important reasons why cybersecurity risk is unique and not like other enterprise risks.

A Q&A session was facilitated by Tony Cole, DDN.QTE and DDN Advisory Board member between Dr. Mussington and DOMINO 24 attendees. Key points from this discussion included:

- The U.S. government views cybersecurity not as a tech issue but a critical issue for our society.
- Enabling *radical transparency* of the system is an objective to counter the evolving nature of risk.
- Risk is not distributed evenly, but their impacts can be.
- There's a chronic information asymmetry problem in place where victims know less than the hackers do, adding to the unique nature of this risk.

Dr. Mussington encouraged the leaders in attendance to engage with their regional CISA representatives to work on these issues to help solve some of these most pressing challenges.

Frameworks: New Developments

No board or director needs to start from scratch on digital, cybersecurity and systemic risk governance. There is a fairly established and proven body of policies and processes in the form of standards, frameworks and models that offer thoughtful and proven guidance reflecting how others are solving these problems.

From the long-established ISO 38500 and ISO 27014 standards to the 2024 update of NIST CSF 2.0 that now includes a *GOVERN* function, to the new disclosure standards imposed by the U.S. Securities and Exchange Commission and the DDN BLAST RADIUS-FALLOUT (BRFO) process for determining cybersecurity incident materiality, the advancing development of these standards is helping boardroom leaders around the world strengthen their processes for digital, cybersecurity and systemic risk governance.

HEARD AT DOMINO

Having your board's audit committee govern cybersecurity is like planning for war where all of your Generals are accountants.

These frameworks reflect the developing maturation of the boardroom and digital and cybersecurity governance as a critical part of the digital business systems that power companies around the world.

Every Board's Starting Point: The ISO Standards

ISO has two well-developed standards focused on governing the digital upside, *ISO 38500 Information Technology — Governance of IT for the Organization* and for protecting the downside, *ISO 27014: Information Security, Cybersecurity, and Privacy Protection — Governance of Information Security.*[iv]

These two foundational standards reflect a basic underpinning of processes, practices and policies in digital and cybersecurity governance that are readily available to any public or private company board.

This guidance comes from the world's leading standard's setter, the International Organization of Standardization (ISO). ISO is an organization that "brings together experts globally to share knowledge and develop voluntary, consensus-based, market relevant international standards that support innovation and provide solutions to global challenges."

The ISO standards are voluntary, and standards can also be in the form of regulatory requirements and leading practices. Overall, there are fewer developed standards focused on governing the board's role in digital value creation than there are for governing the downside of cybersecurity and

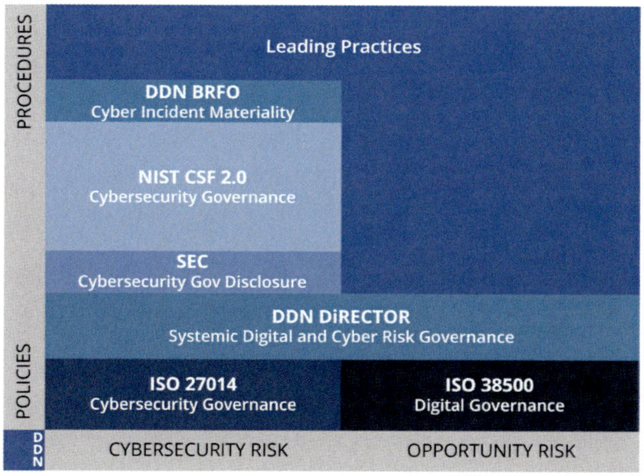

2024 Digital Governance Standards Landscape © DDN LLC

systemic risk. Governing opportunity risk still relies most heavily on boardroom leaders figuring out what works, although the ISO 38500 standard and the DDN DiRECTOR framework can give them a useful head start. There is more on DiRECTOR later in The DOMINO Guide.

ISO 38500 starts by articulating the business benefits of governing IT effectively at the board level including:

- Innovation in services, markets, and business.
- Alignment of IT with business needs.
- Appropriate implementation and operation of IT assets.
- Clarity of responsibility and accountability for both the supply of and demand for IT in achieving the goals of the organization.
- Business continuity and sustainability.
- Efficient allocation of resources.
- Good practice in relationships with stakeholders.
- Actual realization of the expected benefits from each IT investment.

The ISO framework then identifies six key principles to guide behavior and decision making.

The main common-sense tasks these principles address include the evaluation and current and future use of IT, the need to direct the preparation and implementation of strategies and policies to make sure that the use of IT is business aligned, and the need to monitor performance against strategy.

ISO 38500 also recommends policies such as ensuring management capability and responsibility, oversight of emerging technology and its use and implementation, the strategic alignment to business objectives, information, IP and operational risks. ISO 38500 provides any boardroom with a useful reference model for their core corporate governance processes.

With regard to governing the downside risks of complex digital business systems, or cybersecurity, ISO 27014 is the starting point for any corporate board to establish a basic policy benchmark. ISO 27014 establishes the following six key cybersecurity governance objectives:

- Establish integrated comprehensive entity-wide information security.
- Make decisions using a risk-based approach.
- Set the direction of acquisition.
- Ensure conformance with internal and external requirements.
- Foster a security-positive culture.
- Ensure the security performance meets current and future requirements of the entity.

ISO 27014 also identifies four corporate governance processes that should work together to effectively enable the board to oversee information systems. These policies and processes should enable directors to **effectively evaluate, direct, monitor, and communicate** on relevant cybersecurity issues. These four processes reflect a system of proactive actions that corporate directors should perform to govern cybersecurity effectively.

Notably ISO 27014 reinforces the need for corporate directors to be able to evaluate management's information security policies and programs, not just ask questions about them. This would require a level of director cybersecurity expertise not present on most corporate boards at present.

Leading Practice: Benchmark existing boardroom policies and practices against ISO 27014 and ISO 38500. This will provide a way to assess the board's current state of digital and cybersecurity governance policy maturity against a reliable and foundational globally recognized baseline.

NIST CSF 2.0 Affirms The Board As A Key Cybersecurity Control

In February 2024 the National Institute of Standards and Technology (NIST), a part of the US Department of Commerce, released an update to their widely adopted Cybersecurity Framework (CSF) to reflect governance as a key component of the cybersecurity system.

NIST CSF 2.0 is not a regulatory requirement, it is a voluntary standard. It goes beyond just policy to start to reflect practices for governing these issues.

> **HEARD AT DOMINO**
>
> **Frameworks are like golf clubs. You don't use every club on every hole — you use the best club for each particular shot.**

In using CSF 2.0, NIST says:

> **The CSF should be used in conjunction with other resources (e.g., frameworks, standards, guidelines, leading practices) to better manage cyber-security risks and inform the overall management of information and communications technology (ICT) risks at an enterprise level.**[v]

DDN composed a working group during 2023 that providing input to NIST on CSF 2.0.

As a high-level principles-based model, NIST CSF 2.0 is useful for corporate directors to begin to understand the systemic nature of cybersecurity risk and the primary activities and processes associated with effective cyber risk governance and management. Many corporate directors are generally familiar with the first generation of NIST CSF and its use to evaluate the design of a company's cybersecurity program.

NIST CSF 2.0 is organized based upon the core functions intrinsic to a comprehensive approach to cybersecurity. These include IDENTIFY, PROTECT, DETECT, RESPOND, RECOVER and now a sixth function, *GOVERN*.

The addition of GOVERN as a discrete function within NIST CSF 2.0 recognizes the fact that the boardroom is a control in the larger cybersecurity system. As a leadership control, it is a critical control that can positively or negatively impact the performance of the entire cybersecurity system and how it functions to mitigate cyber risk effectively.

NIST CSF 2.0 defines the GOVERN function as follows:

GOVERN (GV) — The organization's cybersecurity risk management strategy, expectations, and policy are established, communicated, and monitored. The GOVERN Function provides outcomes to inform what an organization may do to achieve and prioritize the outcomes of the other five Functions in the context of its mission and stakeholder expectations. Governance activities are critical for incorporating cybersecurity into an organization's broader enterprise risk management (ERM) strategy. GOVERN addresses an understanding of organizational context; the establishment of cybersecurity strategy and cybersecurity supply chain risk management; roles, responsibilities, and authorities; policy; and the oversight of cybersecurity strategy.

There are six categories within the NIST CSF 2.0 GOVERN function:

- Organizational Context
- Risk Management Strategy
- Roles, Responsibilities and Authorities
- Policy
- Oversight
- Cybersecurity Supply Chain Risk Management

Within these categories, there are a total of 31 objectives or leading practices that NIST CSF 2.0 recommends. For example, this practice is included within *Risk Management Strategy* as a key GOVERN policy and leading practice: Calculating the economic impacts of cybersecurity risk is a DDN leading practice and a masterclass was taught at DOMINO 24 on this issue. See the discussion on this masterclass later in The DOMINO Guide.

> A standardized method for calculating, documenting, categorizing, and prioritizing cybersecurity risks is established and communicated. — NIST CSF 2.0

Another notable leading practice included in NIST CSF 2.0 given the CrowdStrike incident of 2024 is included in the *Cybersecurity Supply Chain Risk Management* category:

The NIST CSF 2.0 GOVERN category reflects a comprehensive collection of cybersecurity governance leading practices. While voluntary, leading boards should assess and self-regulate these practices into their boardroom policies, processes and procedures as an important normative tactic to strengthen and mature corporate governance on these issues.

> The risks posed by a supplier, their products and services, and other third parties are understood, recorded, prioritized, assessed, responded to, and monitored over the course of the relationship. — NIST CSF 2.0

During DOMINO 24 Graeme Payne was recognized for his thought leadership in aligning NIST CSF 2.0 with DDN's DiRECTOR framework which is used for identifying systemic risk in complex digital business systems. His paper was selected as one of four thought leadership submissions that will advance the body of knowledge in digital and cybersecurity governance. His full paper is included in the APPENDIX and covers NIST CSF 2.0 in more detail.

Leading Practice: Benchmark existing boardroom cybersecurity governance practices against the 31 practices reflected within the core six categories of the GOVERN function of NIST CSF 2.0. NIST CSF 2.0 is exclusive to cybersecurity risk governance and does not reflect opportunity risk or systemic cyber risk apart from supply chain risk. It provides a more detailed benchmarking standard at the governance procedural level when compared to the ISO standards which are more policy focused.

Calculating The Economic Impacts Of Cyber Risk

Most companies understand and project their estimated losses from doubtful accounts receivables. Manufacturers understand their warranty liabilities and even account for these estimated future losses in their financial statements. Banks also make estimates of loan losses, also reflecting these reserves on their financial statements.

Companies can and are starting to calculate their estimated economic exposures to cybersecurity risk. This is even included in NIST CSF 2.0 as a recommended procedure in the new GOVERN function under Risk Management Strategy as previously highlighted.

Corporate directors cannot understand cybersecurity risk and the organization's overall cyber risk profile until cyber risk is put into the context of economic value at risk. Understanding cybersecurity threats and controls effectiveness is also an important part of the work of the corporate boardroom on these issues, but without knowing what is at stake from an economic perspective, corporate directors will only have a part of the story. **This would be like playing Blackjack in Las Vegas and not knowing the value of the chips you were betting with.**

This issue is growing in importance because digital business systems now drive a majority of business value for many, if not most companies. And companies are largely self-insured for the vast majority of economic exposure from cybersecurity risk.

UnitedHealth Group has declared that the costs related to their 2024 incident could exceed US $2 billion. Business interruption is a favorite objective of malicious attacks because shutting a business down gets hackers paid. But as we saw with the CrowdStrike incident, digital business interruptions are not always created by a malicious attack. And cybersecurity risk or weakness within one company, can impact many companies.

The CEO of Delta Airlines has said that the CrowdStrike incident has cost the airline approximately US $500 million. Delta was forced to cancel 30% of its flight schedule over a period of five days which stranded over 500,000 passengers. From July 15, 2024 until August 1, 2024 CrowdStrike's stock price was down 40% with a class action lawsuit filed within weeks of the incident.

These types of economic implications make cyber risk quantification an important leading practice for any boardroom and management team. The estimated economic impacts of cybersecurity incidents, whether caused by an accidental or malicious occurrence are far ranging and can include revenue generation, profit impairment, negative impacts to employee productivity, product quality, supply chain management, litigation, recovery costs, share price declines and many others.

Every corporate board and management team should have a deep understanding of the amount of economic exposure the company faces as a result of the risks to or within their digital business systems.

Kevin Richards, President of Cyber Risk Solutions at X-Analytics taught a masterclass on board reporting and estimating the economic impacts of cyber risk at DOMINO 24. Covering an actual case study of this leading practice, he demonstrated how transformational this practice is in improving how a board and management team approach cybersecurity risk mitigation activities including risk transfer through cyber insurance.

> Understanding the economic impacts of cyber risk helps boards make informed decisions that protect their company's bottom line and stakeholder interests.
>
> —Kevin Richards, President X-Analytics

Reflecting on his actual experience with a high-profile cybersecurity incident, Kevin's case study demonstrated how useful and effective a well thought out and calculated determination of the estimated economic exposures from a cybersecurity incident are. He said this was particularly useful in two ways. First, in designing and executing an incident response plan as the decision makers could calibrate to a thoughtfully determined reference value as the incident unfolded. Second, it changed how the company considered risk mitigation strategies from the controls they had or needed that self-insured this exposure, to how much quantifiable risk they wanted to transfer through cyber insurance.

Knowing this allowed the board and management teams to have contextual discussions when needed with investors and other stakeholders before and throughout the incident. .

Emphasizing the fundamental fact that every company is largely self-insured for the economic exposures of cyber risk (DDN estimates that 80-85% of the economic exposure from cyber risk is self-insured), Kevin's masterclass put the issue into the strategic and tactical contexts boards and management teams need to effectively manage and mitigate these risks.

Also addressed in his masterclass was the need for board reporting to align cybersecurity exposure to the primary economic loss categories used by the cybersecurity insurance market of:

- Business interruption
- Data breach
- Ransomware
- Asset misappropriation

By translating these possible impacts into the context of the organization's key financial statement metrics such as revenue and profitability, balance sheet assets, cash flow and market capitalization, boards will have a contextual understanding of their overall cybersecurity risk profile.

Risk appetite is all about understanding what is at stake economically and risks tolerances relate to what the risk levels are to that stake. The two of these together determine an overall risk profile for a particular desired outcome.

NIST CSF 2.0 also makes this leading practice recommendation in the new GOVERN category within the Risk Management Strategy area:

> — Risk appetite and risk tolerance statements are established, communicated, and maitained.

Kevin summarized several leading practices for determining the economic impacts of cybersecurity risk and communicating within the boardroom:

- Reframe cybersecurity risk around the financial context of *economic value at risk* including projected impacts on key financial statement items and measures.
- CISO's should be the champions of this approach and have a structured process to communicate, educate and evangelize this issue with management including the CEO and CFO and the full board.
- Adopting a process to do this also helps with the overall qualitative and quantitative analysis of cybersecurity incidents for SEC disclosure purposes, a process CISO's should own.

Leading Practice: Ensure the management team has estimated the economic loss exposures related to cybersecurity and systemic cyber risk. These determinations will change as the company's digital footprint evolves and the risk environment changes, so these exposures should be regularly monitored. Boards can and should also engage third-party experts to quantify these exposures, or to verify management's determinations. Boards should also understand how much of this exposure is covered by cyber-insurance versus self-insured. The company's premium/coverage ratios for cyber insurance versus self-insurance is an exceptionally useful metric for boards to understand.

The New SEC Cybersecurity Disclosure Rules

Regulations are forced standards that coerce companies who are bound by the regulation to do what they were not or did not want to do. Regulators play an important role in standards development and adoption as they step in when change needs to be forced on a particular issue. Coercion through regulation often drives widespread new behaviors in a short time frame.

But government regulators typically lag the reality of market risks and government mandate for cybersecurity governance is no exception.

However the U.S. Securities and Exchange Commission recently passed new cybersecurity incident, management and governance disclosure standards that went into effect in December 2023 for most companies. More policy than procedure, these new disclosure rules are transformative and will force beneficial change in the boardroom and with management teams on their cybersecurity policies and practices.

Over the last decade, there have been a significant number of new cybersecurity and data privacy regulations passed by governments around the world and more is coming, e.g., GDPR, CCPA, DORA. Management teams have had to comply with these many rules and boards have had to ensure managements adherence to them. But regulation that impacts the boardroom and governing cybersecurity risk has been almost non-existent anywhere around the world.

The SEC has now changed that, but at the same time created a significant amount of boardroom and management confusion with their new SEC cybersecurity disclosure rules. A *Cheat Sheet* summarizing the requirements of the new rules is included in the APPENDIX of The DOMINO Guide.

This confusion is actually a sign that these regulations were needed. It demonstrates the lack of process maturity and the level of underdeveloped management understanding surrounding cybersecurity given the struggles in applying them in the context that the SEC imposed.

> DDN has written extensively on the new SEC disclosure rules and the struggles with their implementation. DDN has even created the only masterclass/tabletop and process for making a cybersecurity incident materiality determination pursuant to the new SEC rules. This is addressed later in The DOMINO Guide and was taught at DOMINO 24 in a masterclass by Kudelski Security.

Those who attended DOMINO 24 had the record set straight with the new rules with some legal depth and insight provided in a masterclass taught by Maia Gez, Partner and Head of the Public Company Advisory Group at global law firm White & Case.

In Maia's masterclass, she kicked off by covering the key fundamentals of U.S. securities law for cybersecurity disclosure including the long-standing U.S. case law based concept of materiality for disclosure purposes. She also clarified that this is not the same concept of materiality as frequently applied for financial reporting or auditing purposes which is a common misunderstanding that has contributed to the implementation confusion.

Notwithstanding these initial implementation hurdles, the SEC's cybersecurity disclosure rules are transformational, and their intent is simple:

Disclosure is a powerful means to an end for the SEC — and that end is to get companies to mature their processes and understanding of the cybersecurity risks facing their businesses from an investor's perspective. **The final rules are not prescriptive in terms of dictating policies and procedures in cybersecurity governance and management — the SEC left the procedural details to management teams to figure out.**

> ...to enhance and standardize disclosures regarding cybersecurity risk management, strategy, governance and incidents by public companies that are subject to SEC reporting requirements. — SEC Final Rules

Acknowledging that their own guidance in 2011 and 2018 did not meet its intended objectives, the new SEC disclosure rules are a landmark step forward that forces focus and action on key cybersecurity policies and procedures at the management and board level for U.S. public companies.

CISO's should view the new SEC disclosure rules as an opportunity and the arrival of a leadership moment. The boardroom and C-suite will look to the CISO for interpretation, guidance and leadership in implementing them. And corporate directors should view them as a first regulatory step towards being held accountable to a higher standard relative to their role in cybersecurity.

Applying the new cybersecurity disclosure rules requires companies to understand, explain and develop cybersecurity processes now subject to SEC disclosure — often for the first time. Some of these processes may be non-existent or not very mature or well documented. Other management team members and those involved with disclosure processes now need to understand cybersecurity and how the digital business system drives value for the organization to a greater extent than they may have understood previously — which is exactly what the SEC wants companies to improve upon.

The SEC's enforcement division is monitoring the new disclosures closely and being cooperative with issuers to get through some of the initial implementation challenges.

A webinar with DDN's Bob Zukis and SEC Corporate Finance Director Eric Gerding is available on the DDN LinkedIn corporate site. This discussion addresses some of the market issues and implementation challenges.

Advice from Maia Gez, Partner at White & Case during her masterclass on implementing the new SEC disclosure rules included:

Disclosure Controls and Procedures

- Ensure there are appropriate procedures for reporting and escalating incidents to the disclosure committee for materiality determinations.
- Revisit processes for evaluating materiality of cybersecurity incidents and consider which "elevation criteria" are used to identify the incidents elevated to the disclosure committee for materiality assessments.
- Disclosures will need to be considered and prepared while the breach is evaluated, so clear processes, coordination and consistent documentation are necessary.

Cybersecurity Policies and Procedures, Including Third-Parties

- Evaluate, or audit the adequacy of existing cybersecurity policies and procedures.
- Integrate adequate disclosure controls and procedures, including 8-K reporting/timing into cyber incident response plans.
- Identify third-party service providers and review contractual provisions and communication protocols to facilitate timely assessment and disclosure of incidents.

Review And Assess Governance and Oversight Structure

- Reassess board and management oversight

processes and consider improvements such as scheduling additional cybersecurity updates on meeting agendas and committee alignment/ structure.

- Consider increasing the time spent addressing cybersecurity and strengthening processes for timely communications and documentation internally.
- Evaluate whether the board and management have appropriate cybersecurity expertise.
- Enhance board education on cybersecurity and view cyber literacy as similar to financial literacy with a goal to ensuring all board members have a sufficient basic understanding.

Re-Assess Annual Cybersecurity Risk Disclosures

- Assess for scope, clarity and consistency the existing disclosures across annual report risk factors and business section, proxy statement and sustainability report.
- Prepare new disclosures pursuant to new requirements including role and experience of management, persons responsible for cybersecurity and processes for identifying material incidents, addressing systemic cyber risk and particularly third-party risk.

Maia emphasized her comments about the importance of consistency with documentation and focusing on the process. Maia emphasized the legal exposures particularly around a disclosure that says you are doing something, e.g., a particular process or procedure, when in reality you are not.

Leading Practice: Monitor global regulatory developments from regulatory bodies around the world specific to corporate governance policy and procedure given the increasing amount of focus on the role of the board in cybersecurity. Alongside this, apply a boardroom continuous improvement model by monitoring new practices from boards around the world that can optimize boardroom effectiveness as new policies and procedures emerge and mature.

Determining Cybersecurity Incident Materiality

Another headline surrounding the SEC disclosure rules has focused on the confusion surrounding the implementation of the cybersecurity incident materiality disclosure requirements. Particularly, what is materiality and how to determine it when there is a cybersecurity incident.

Regardless of this only applying to SEC covered companies (both U.S. and Foreign Private Issuers), it is a process that every company should have, whether private or public, U.S. or not. Being able to understand a cybersecurity incident and it's far reaching qualitative and quantitative impacts is necessary for every boardroom and management team.

A masterclass/tabletop was held at DOMINO 24 on just that issue. Kudelski Security's Michael Barnes and DDN Faculty and DDN.QTE Jason Fruge taught a masterclass on DDN's BLAST RADIUS – FALLOUT (BRFO™) process. This process fulfills the SEC's expectations for companies to have an *informed and deliberate process* for making a cybersecurity incident materiality determination and for identifying what to disclose. Several case studies were covered in this highly engaging session. Key points covered included:

- The BRFO process reflects the need to understand and disclose the material **aspects** and **impacts** of a cybersecurity incident.
- While applicable to cybersecurity incidents caused by malicious intent, the definition of cybersecurity incident also includes accidental occurrences that impact the confidentiality, integrity or availability of information systems used or owned by an SEC registrant.

While this may sound straight forward, it is the application of the SEC's securities case law based definition of materiality within the context of a reasonable investor and the far-ranging impacts of a cybersecurity incident where the implementation challenges are arising.

Requiring both a qualitative and quantitative approach to identifying the material impacts of any cybersecurity incident, the SEC has put the burden of this determination on the issuer. This disclosure requirement will force companies to understand the far-reaching aspects of an incident and how the incident can impact stakeholders inside or outside of the company both qualitatively or quantitatively.

This is exactly what the SEC wants management teams to critically think about — and have a process for.

The objective nature of this type of analysis and the requirement to have a structured approach are new challenges for many management teams. Because of this disclosure requirement, a broader group of executives are also having to understand cybersecurity and how the digital business system drives value to a greater extent than they have previously had to. That again, is a good thing. The BLAST RADIUS-FALLOUT Process (BRFO™) can help boards and management teams fulfill these new requirements.

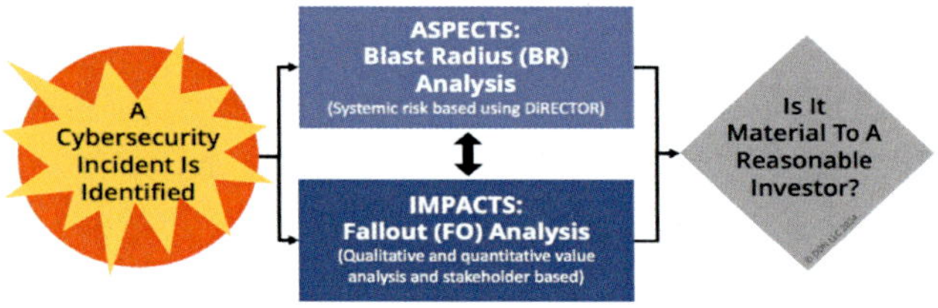

The DDN BRFO™ Process © DDN LLC

The goal of the BRFO process is to reflect what the SEC expectation is for this disclosure requirement — for issuers to have an informed and deliberative process in place for making this determination. Key benefits of BRFO are:

- It is quickly and easily implementable.
- It is repeatable.
- It documents key issues in analyzing any incident consistently.
- It considers systemic cyber risk in all aspects related to an incident.
- It covers both qualitative and quantitative impacts.
- It is universal to any company, sector or incident.
- Using it creates a disclosable process for incident materiality identification.
- It can be used if subject to SEC rules, or not.

A detailed flowchart of the BRFO process is available for DDN members online and private masterclass/tabletops can be arranged for individual companies.

As part of the SEC's new disclosure rules for management and governance processes, the SEC also wants issuers to annually describe their:

> ...processes, if any, for assessing, identifying, and managing material risks from cybersecurity threats in sufficient detail for a reasonable investor to understand those processes.

The BRFO process is an effective and disclosable process that can meet the SEC expectation for how management teams identify material risks from any cybersecurity incident.

Leading Practice: Ensure that management has an *informed and deliberative process* in place for making a cybersecurity incident materiality determination. This includes clearly established accountabilities for administering it, alignment with disclosure processes, consistent documentation approved by legal counsel, and incident response team training. Conducting an annual cross-functional tabletop exercise focused on cybersecurity incident to materiality determination with corporate director participation is also a strong leading practice.

The SEC has disclosed that their enforcement agenda for 2024, and likely longer, includes a focus on information security and operational resiliency. They emphasized this by stating that their enforcement agenda will:

> ..focus on registrants' policies and procedures, internal controls, oversight of third-party vendors (where applicable), governance practices, responses to cyber-related incidents, and training.

HEARD AT DOMINO

The SEC is adopting a collaborative approach on the rollout of the new rules, so far. Their goal is not to second guess decisions on materiality, but to ensure issuers are focused on having an effective process, consistent documentation and critical thinking. It's really not a right or wrong gotcha situation, it's about the process.

Heightened regulatory mandated disclosure and scrutiny is intended to focus companies on maturing their processes on cybersecurity risk management and governance.

While a powerful coercive force of change in the short term, the new SEC disclosure rules do not go as far as some companies and boardrooms already do from a leading practices perspective. Leading practices in both cybersecurity governance and management already extend well beyond the SEC's new cybersecurity disclosure rules.

Leading Practice: Complete a full benchmarking and grading of the board's system of governance against leading practices covering director skills, boardroom structure and scope of risk oversight. This benchmarking should be against all of the company's public facing statements and disclosures to assess completeness and alignment of investor communications, e.g., 10-K, 10-Q, Proxy, committee charters, etc.

Risk Management Regulation Updates From Europe

Erwin Rezelman, DDN.QTE and SVP, Head of Technology for Continental Europe at global insurer SOMPO delivered an update on several key European regulations. While these are not all prescriptive of specific boardroom policies, practices and procedures, they are very governance relevant as indicators of the regulatory path forward for the digital economy. These two updates reflect regulations focused on digital risk management and operational resiliency that will require boardrooms to engage on these new rules. These regulations can be viewed as additions to the standards base.

DORA

HEARD AT DOMINO

The SEC doesn't want companies to just ring the alarm bell on a cybersecurity incident with disclosure, they want the company to describe to investors why they felt the need to ring the bell in the first place.

The EU Digital Operational Resilience Act (DORA) establishes an Information and Communication Technology (ICT) risk management framework for the EU financial sector. It sets technical standards for financial entities and their third-party technology service providers to ensure operational resilience in response to ICT related disruptions and threats. It will go into force in January 2025.

DORA governs all financial services companies with a license to write business in the EU. It also applies to ICT services providers of financial institutions in the EU. The ICT services provider can be based anywhere in the world.

For financial institutions the requirements are structured around five key ares:

- ICT risk management.
- Cyber incident reporting and recovery.
- Operational resilience testing.
- Third party risk management.
- Reporting and information sharing.

As part of DORA, the financial institutions will need to create a registry of information of all its ICT vendors and report major incidents to the regulator within 24 hours, with follow-up reports. This unique regulation will force companies to identify their critical systemic cyber risk landscape of third-party risk within their digital business ecosystem.

ICT Vendors will need to maintain an office in the EU, adopt the DORA clause as part of the service contract with an EU based Financial Services company, and immediately report major incidents to its customer(s).

The penalty for non-compliance can go up to 2% of the total worldwide revenue of the financial institution. Whereas for critical third-party ICT service providers, the fines can amount to Euro 5 million and individuals in these companies may face fines of up to Euro 500,000.

Increasing individual accountability is a notable emerging trend in cybersecurity in the U.S., Europe and Australia.

EU Artificial Intelligence (AI) Act

The EU is also ahead of most other regulators in understanding that Artificial Intelligence (AI) technologies create new risks with far ranging impacts that need new approaches to risk management. And they are much faster at forcing companies to govern and manage it through regulation.

Also risk based, the EU AI Act will come into force over the next several years.

The EU AI Act is a risk-based regulatory framework to manage the development and use of AI to ensure the technologies align with European values, including fundamental rights, safety and transparency. It is one of first collections of overarching governance principles put into law focused on the application and use of AI with the aim to balance innovation and the need to protect EU citizens.

Like DORA it is risk based and categorizes AI risk into four risk types with different requirements based upon levels of risk:

- Unacceptable risk. AI systems considered harmful to fundamental rights and/or ethical standards.
- High Risk. AI that significantly impacts user's lives or rights, including safety.
- Specific Transparency Risk. To foster trust, AI that poses some risk that needs transparency.
- Minimal Risk. AI that covers most consumer applications and will be unregulated under this law, but subject to existing laws.

There are transparency and human oversight requirements, especially for the first two categories. The penalties for non-compliance are up to 6% of the company's annual revenue or Euro 30 million, whichever is higher.

These EU regulations will force management teams to respond by further developing their understanding of risks related to these technologies and by being more transparent and accountable on their impacts on behalf of their stakeholders.

Both of these regulations are risk based, and focused on material impacts of the technology and cybersecurity risk that will put new requirements on boardrooms and management that will force them to understand these technologies and their impacts and risks. While not yet imposing specific processes or procedures for corporate boards, they will nonetheless raise the bar for corporate directors in executing their fundamental duties and understanding the digital risk environments facing their companies.

EUROPEAN CENTRAL BANK

The ECB has imposed the first known rules specifically addressing director cyber expertise and the ability of the boardroom to effectively oversee cybersecurity risk. **As a policy they expect the banks they oversee will have at least one non-executive board member with recent practical expertise in Information and Communication Technology (ICT) risks and security risks.** They suggest that five years of practical experience is an adequate threshold for fulfilling the expertise requirement.

The global regulatory trends are focused on risk, transparency and accountability. Risk is about forcing companies to have a better understanding of the impacts of cybersecurity incidents across their stakeholders. Transparency relates to improving their understanding of the details of these technologies. Accountability is heavily focused on management responsibility, but the desire for greater boardroom accountability has also started, and is inevitable.

2. The System Of Digital And Cybersecurity Governance: Skills, Structure And Scope

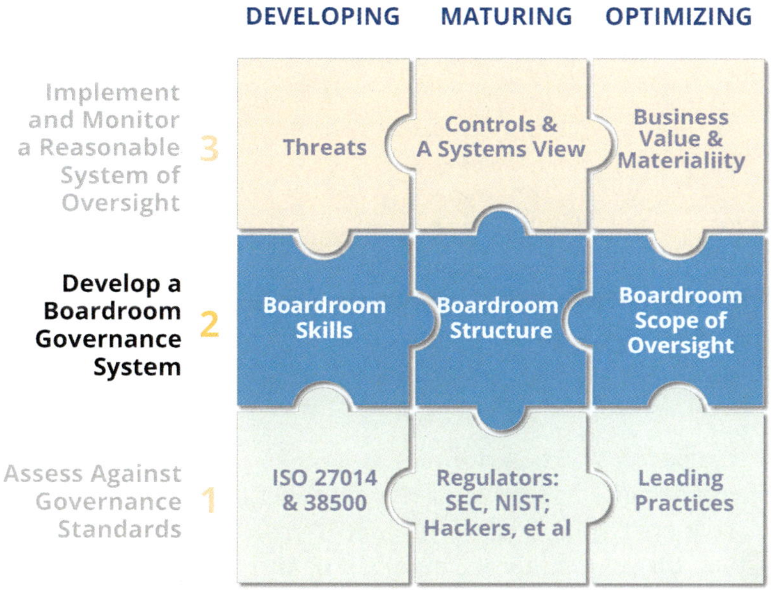

The key to solving this puzzle is to take a systems view to the board's policies and procedures to create an adaptive and high-performing system capable of governing the digital opportunities and risks of today, and tomorrow.

Comprised of three parts, this governance system reflects the right director *skills*, an effective boardroom organizing *structure*, and a comprehensive approach to governing the *scope* of risk oversight surrounding the complex digital business system. A coordinated collection of policies, processes and capabilities in each of these areas needs to work together to enable the boardroom to be adaptive to this rapidly changing environment, while remaining effective.

Corporate boards are constantly presented with new challenges in governing transformative new technologies, e.g., AI, quantum computing, digital twins, augmented reality, systemic cyber risk, etc. Every board has a responsibility to both understand the strategic implications of these tools alongside their new risks. These risks can also change the overall risk profile of the organization's existing digital business environment.

The approach that boards take to respond to the scale, scope, and pace of change in IT cannot be left to audit committees and directors who lack the expertise to understand these technologies. Boards cannot react to new developments with ad-hoc changes in boardroom process or policy or exclusively rely on outside "experts". The boards approach to effectively govern these dynamics requires a high performing and adaptive system in and of itself within the boardroom.

Boards need to solve these challenges by thinking and acting systemically. There is not a single policy or process that can solve these challenges. The board is a key part of the organization's complex digital business system, both with regard to how it creates value and how that value is protected. As a leadership control, the boardroom has the ability to make the entire system stronger if high-performing, or weaker if low, or non-performing.

A systems view on these issues is not an entirely new concept in the boardroom. It is actually ingrained in the significant corporate governance reforms imposed by the Sarbanes-Oxley Act of 2002 related to strengthening the board as a control in the financial reporting system — and it works.

Skills: Digital, Cybersecurity, And Systemic Risk Expertise In The Boardroom

Empirical evidence is growing which shows that significant positive business benefits are realized by companies that have corporate directors with true digital, cybersecurity and systemic risk expertise on the board.

Just as boardrooms needed qualified financial experts to govern financial reporting effectively, boards need qualified technology experts to govern the digital business system effectively — and this expertise extends beyond directors with cybersecurity expertise.

HEARD AT DOMINO

Plain and simple — you can't effectively govern what you don't understand. Every board needs directors with expertise across the complex digital business system in multiple domains from emerging technology to IT operations. Director cybersecurity expertise is just table stakes, and how is that even debatable?

As boards develop a high-performing system of digital, cybersecurity and systemic risk oversight, director *skills* are the most powerful control within that system. **Director skills are a powerful and irreplaceable control because boardroom leadership makes the entire system stronger.**

Research from MIT has identified the significant business benefits that are achieved when a board has a critical mass of digitally savvy directors as addressed previously. This critical mass reflects the breadth of expertise needed to understand the complexity of the entire digital business system. The business benefits identified by MIT include higher revenue growth, profitability, return on assets, and market valuation.[x]

Evidence also shows the positive business benefits when boards have directors with cybersecurity expertise, and the negative impacts when they do not. Research from Virginia Tech about the presence of director cybersecurity expertise on the board concludes that value is added to the overall cybersecurity position of the organization that is not possible without director cybersecurity expertise. Furthermore, they identified that without director cybersecurity expertise boards have a *check-the-box* approach to cybersecurity.

These issues were discussed at DOMINO 24 from both sides of the boardroom table in peer learning sessions with a group of public company corporate directors facilitated by E&Y and Proofpoint.

While the SEC dropped the requirement to disclose director cybersecurity expertise from their final disclosure rules, which was a disclosure item that would have likely coerced boards to add director cybersecurity expertise to the board, this remains a very strong leading practice.

Leading Practice: Add directors with true digital, cybersecurity and systemic cyber risk expertise to the board and make more informed and useful disclosure of these director skills in proxy statements. Adopt the DiRECTOR framework as a director competency model for internal reference to understand the director capabilities needed to govern the entire complex digital business system.

Analyzing the comment letters that were submitted to the SEC on director cybersecurity expertise shows two very different groups working FOR and AGAINST this issue.

The APPENDIX reflects a detailed write-up and analysis of both the FOR and AGAINST camps and their rationales as presented in their comment letters to the SEC.

The arguments in the AGAINST camp superficial and illogical, while the FOR camp is comprised of those with deep cybersecurity expertise who also brought empirical evidence surrounding the positive business impacts to validate their support. The FOR camp is comprised of organizations with deep cybersecurity expertise and leadership, institutional investors, academics who researched the issue, the AICPA and government regulators.

The AGAINST camp was comprised of two national corporate governance associations, the American Bar Association and several law firms, several accounting firms, a large group of industry/trade associations who are parroting the U.S. Chamber of Commerce, and Microsoft.

The AGAINST camp would also implicitly include hackers, as their interests are served by keeping leadership weak on cybersecurity. Presenting no evidence for their arguments, there was also a coordinated negative propaganda campaign identified in the AGAINST camp reflecting a collection of alarmist and unsupported fallacies.

The SEC also rationalized the omission of director cybersecurity disclosure from their final rules with the statement:

'We are persuaded that effective cybersecurity processes are designed and administered largely at the management level, and that directors with broad-based skills in risk management and strategy often effectively oversee management's efforts without specific subject matter expertise..."

We believe this statement shows a lack of understanding of not only the unique nature of cybersecurity risk including its systemic aspects and impacts, but also ignores an important SEC precedent. **The SEC has previously forced boards to disclose director financial expertise through the Sarbanes-Oxley regulations with the goal of strengthening the board as part of the financial reporting system.**

The European Central Bank is the first known regulatory body that taken this step for the banks under their supervision, with an expectation that boards will have at least one non-executive director with recent and practical Information and Communication Technology (ICT) and security risk expertise.

HEARD AT DOMINO

Directors aren't there to just ask questions as questions are useless if you don't understand the quality of management's answers. That's the goal behind digital and cybersecurity expertise in the boardroom and that's what effective directorship is, or should be, about.

DDN believes this omission was a significant mistake on the part of the SEC that has not only contributed to the chronic cybersecurity weakness of America's companies but exacerbated them.

DDN POLICY ALERT: For DDN a director with cybersecurity expertise on the board is a key policy position we strongly advocate for. We have made an open call to all U.S. public companies to add director cyber expertise to their board as this is the starting point for strengthening the board as a control in the cybersecurity system. Read the DDN key policy positions in The DOMINO Guide APPENDIX.

Leading Practice: Deliver annual director training on each aspect of risk, i.e., opportunity, cybersecurity and systemic cyber risk to develop a base level of competency for all corporate directors on these issues.

Structure: Organizing The Boardroom For Optimal Oversight

How the board organizes itself to execute its responsibilities plays a significant role in how effective corporate directors are in overseeing an issue.

Recent cybersecurity incidents at Crowdstrike, Clorox, MGM, UnitedHealth Group and Caesars all have one thing in common — boardroom audit committees with responsibility for governing cybersecurity.

This is a common bad practice. But the leading practice of moving cybersecurity oversight out from underneath the audit committee and into a focused committee is already transforming over 200 of America's R3000 boardrooms, including FedEx, GM, Salesforce, Verisign, Jabil, Hasbro, Verizon, and many other well-known companies. What do these boards understand that others do not?

Research on the benefits of committees within the boardroom identify the following advantages:

- Task efficiency within the committee.
- More time spent on the issue in greater depth and breadth.
- Knowledge specialization within the committee.
- Greater accountability around the issue to internal and external stakeholders.

DDN POLICY ALERT: In 2018 DDN called on all U.S. public companies to add a Technology and Cybersecurity Committee (TCC) to their board as a leading practice to mature digital and cybersecurity governance effectiveness.

The SEC's chief accountant has even questioned whether cybersecurity belongs in the audit committee. Challenging whether the right director skills and attention are able to be provided to the dynamic nature of cyber risk when this is relegated to the busy agenda of the audit committee.

Audit committees themselves are an artifact of the Sarbanes-Oxley regulations. This boardroom organizing principal was a requirement of SOX together with the disclosure requirement for director financial expertise. These were two critical systemic cornerstones for improving the corporate governance system responsible for financial reporting.

The next step beyond removing cybersecurity from the audit committee is to align digital innovation with cybersecurity risk within the same committee, i.e., a Technology and Cybersecurity Committee (TCC). This is also a DDN policy position and recommended leading practice.

Evidence from those with directors with digital and cybersecurity expertise who have experienced this change, some of which was shared at DOMINO 24, supports the transformative positive impacts of having a Technology and Cybersecurity Committee on the board both from a governance and management perspective.

The FedEx board was highlighted at DOMINO 24 as having had a Cyber and Technology Committee in place since 2000. One of the responsibilities of the FedEx directors on this committee as articulated in their Charter is to:

- Advise the Company's senior Data and Technology management team.
- Stay informed of, assess, and advise the Company's senior Data and Technology management team with respect to trends, emerging technologies, applications, and systems that relate to or affect the Company's cyber and technology strategy or programs.[xiii]

TCC's are powerful parts of establishing an adaptive governance model for new information technology developments and innovations, e.g., AI. Given the broadening scope and reach of information technologies, these technologies will only become more paramount in their effects on the organization, easily eclipsing the narrow scope, focus and capabilities of an audit committee or the time availability and focus of the full board.

HEARD AT DOMINO

Investors should look for the presence of a technology and/or cybersecurity committee on the board as an indication of the board's maturity in governing these risks. If it does not exist, they should challenge why cybersecurity is within the audit committee and understand how much time and focus this issue is given, and by whom.

A TCC creates the dedicated focus, imposes the needed time commitment and aligns and applies director expertise to establish the boardroom as an adaptive and high-performing leadership control in the digital business system. DDN has created a leading-practices TCC charter available for DDN members online.

Leading Practice: Implement a Technology and Cybersecurity Committee (TCC) on the board as a leading practice. This will increase boardroom task efficiency, accountability and align director knowledge specialization to these complex issues. This will also move cybersecurity out from the audit committee and align it with digital innovation to create greater boardroom efficiencies and effectiveness.

3. The Scope Of The Oversight System: Opportunity, Cybersecurity And Systemic Risk

Vice Chancellor Glasscock of the Delaware Chancery Court has said that "the evaluation of business risk, [is] the quintessential board function."[xiv]

What is the scope of a reasonable system of oversight regarding the digital business system? What are the core aspects of risk related to complex digital business systems that directors need to govern?

The third part of DOMINO's Puzzle requires a proper understanding of risk related to the complex digital business systems that power a growing amount of corporate value. This is a critical part of strengthening the boardroom, and a prerequisite to solving the challenges of governing digital risk.

The legal responsibilities of corporate boards and directors are familiar to many executives and directors in the phrases and concepts of *duty of care* and *duty of loyalty*. In common parlance in the U.S., the board's responsibilities are articulated in the SolarWinds shareholder derivative lawsuit — to "implement a reasonable *system* of oversight, and then to monitor it".

A big reason behind the chronic failures that persist in cybersecurity resides within the failure of boards and management teams to understand the changing nature of risk brought about by the evolution of digital business systems.

There are three key aspects of risk that define the scope of the board's responsibilities related to digital business systems. The three unique aspects of risk that directors have a responsibility to govern with digital business systems are **opportunity risk, cybersecurity risk and systemic risk**. Actively governing each of these aspects of risk is the final piece of the solution in strengthening the boardroom as a control in the digital business system.

- **Opportunity Risk:** The strategic and value-creating risks of digital technologies that drive economic output and growth.
- **Cybersecurity Risk:** The active and malicious cyber threat environment to the organization and its ecosystem because of the digital business system.
- **Systemic Cyber Risk:** The inherent risks within and between the parts of the complex digital system that threaten the purpose of the system within the enterprise, and beyond.

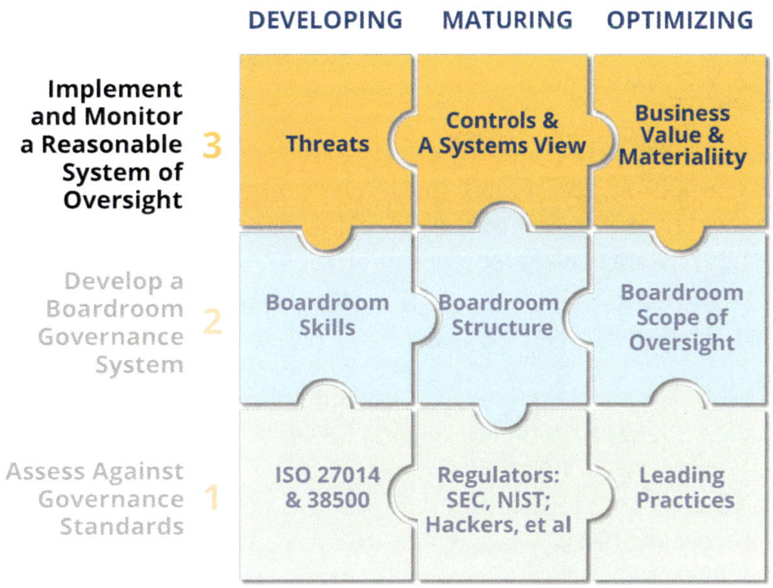

Digital, cybersecurity and systemic risks are unique aspects of risk that are different than what many corporate directors are familiar with. These risks are not generic or capable of being governed with a generalist understanding of risk management.

Just as effectively governing financial reporting needed director financial expertise in the boardroom, the digital business system also needs relevant digital expertise to oversee these unique risks.

Risk is always contextual to the domain it originates from. Understanding risk can require depth of knowledge within the domain that the risk emanates from in order to govern it effectively. Boards frequently make the mistake of believing that audit committees can effectively oversee cybersecurity risk for example. This misaligns director skills, subverts the cybersecurity agenda to the financial reporting oversight agenda and does not bring the proper understanding of risk to the issues related to the digital business system.

HEARD AT DOMINO

Going to your audit committee for cybersecurity risk oversight is like going to your plumber for dental work — those are two very different risk environments. They each need deep and relevant domain expertise to understand risk within each domain, and those capabilities are hardly interchangeable.

The CISO's, CIO's, and corporate director's at DOMINO 24 learned about the scope and challenges of risk oversight as it relates to the digital business system that powers their companies.

Digital Opportunity Risk

The amount of economic growth and output created by, and dependent upon complex digital business systems is material to the economic success of economies and companies around the world. Some companies are digital natives, their entire business is a complex digital business system. For other companies digital business systems can drive much of their economic output and growth including significant contributions to revenue and profitability. Automation, cashless payments, digital products and services, ecommerce, artificial intelligence and employee connectivity are all digital value drivers that can reach new markets, grow revenue, cut costs and improve productivity.

Corporate boards have a responsibility to oversee this digital upside. The ISO 38500 framework addressed previously is focused on these strategic and digital growth responsibilities for the corporate boardroom. Directors need to understand the digital technologies that are driving their company's future, their impacts on strategy and the organization's tactics for competing. This requires much more than a casual awareness of these tools and a periodic boardroom agenda item related to digital innovation and disruption.

Research from MIT shows that when boards are digitally savvy and have the expertise to understand these technologies and the forces they capture and leverage, it drives large amounts of relative business value. Digital leadership in the boardroom matters for business success.

Cybersecurity And Systemic Cyber Risk

DOMINO 24 brought together boardroom and executive leaders who are governing all of these aspects of risk. Patrick Joyce, Resident Global CISO at Proofpoint and former CSO at Medtronic led a learning panel of CISO's who educated their peers about these issues. Information security leaders and DDN QTE's from Stanley Black & Decker, Zebra Technologies and Delta Airlines shared their insights.

A DDN.QTE and CISO at a public company put it simply when she said, "So much strength comes from being brilliant with the basics in cybersecurity." Her comments reflect the need to build a culture of cyber and systemic resiliency throughout the company, an undertaking that highlights the need for boardroom leadership in setting the cyber tone-at-top of the organization.

Brad Skibitzki, DDN.QTE and CISO at Zebra Technologies emphasized that every individual is a unique part of the digital business system which dictates the need to build a culture of cybersecurity awareness and effectiveness throughout the organization. Or in other words, a high-performing system of cybersecurity with the board being a critical part of that system.

Veracode's CISO Sohail Iqbal also taught a masterclass on AI and application security — the very risk that befell SolarWinds. In his masterclass on how application risk creates systemic challenges deep within the digital business system, Sohail discussed how artificial intelligence (AI) tools are creating even more risks that are deeply embedded within the software environment of the complex digital business system.

Highlighting that software vulnerabilities are one of the top three exploited vulnerabilities by hackers that grew 180% in 2023, his masterclass emphasized the point that software security flaws are being found and exploited much faster than they can be fixed.

Exacerbating this problem is the fact the Generative AI is being used to write software code by 92% of developers. With a desire to generate code faster, the tradeoff is software code that is becoming more intrinsically flawed and open to exploitation. AI generated software code also often includes known security vulnerabilities within the AI generated code base, with data demonstrating that this happens over 40% of the time.

This requires a solution that fixes software code as quickly as it is generated which is a problem that Veracode is solving by utilizing these very AI technologies to achieve this. His recommendations for corporate directors on this issue include:

- Ensure risk management programs for third-party software developers monitor the use of Generative AI for software development.
- Monitor how AI is being used in software development, along with other processes, including the integrity levels of the large language models (LLM's) being used by the AI tools for software development and beyond.

Risk isn't static as new innovations often create new threats and new types of risk. These new risks are commonly unfamiliar to those without experience or an understanding of the related technology To believe that cybersecurity risk can be governed and managed within the context of general risk management competencies and perspectives is misguided.

This is something the cyber insurance industry has discovered the hard way with higher- than-expected loss ratios. Corporate boards and management teams need new capabilities and perspectives to understand and manage these distinctive risks.

And the world of risk is expanding. CISO's already had a very difficult job addressing a rapidly evolving threat environment with active adversaries. Systemic cyber risk is now compounding that difficulty as these risks exist inherently within every complex digital system and they are expanding with more impactful consequences. Systemic weaknesses are also increasingly being targeted and leveraged by hackers with wide ranging impact and positive effect — for them.

SolarWinds is an example of a systemic attack where the objective was to efficiently reach SolarWinds client base comprised of over ten thousand customers — through SolarWinds.

Systemically important organizations, such as major cloud software providers, can distribute risk quickly and efficiently causing significant levels of disruption.

Cybersecurity risk reflects the active threats to a business as a result of the digital business system. Systemic cyber risk is about the risks inherent within complex digital business systems whether there is an active adversary or not. Every board has a responsibility to understand and govern both of these issues.

The CrowdStrike incident is an example of the far-ranging systemic risks inherent within the complex digital business systems and the inter-connected nature of this risk environment. That incident was not caused by an attack but by several fairly simple errors that were exported to third-parties.

Companies exist in a world of distributed cyber risk, but there is not a collective security model or leadership mindset that exists to address the systemic nature of this unique risk environment.

Systemic risk in cybersecurity has largely been ungoverned within the boardroom, and frequently underestimated within many management teams. A failure to understand risk in the right context, i.e., to understand the problem, is contributing to the chronic and expanding cybersecurity weaknesses in American businesses, and in companies around the world.

Levels of systemic risk are as high as they have ever been. This has been brought about by the irreplaceability of certain parts and vendors within the system. Issues such as the interconnectedness of systems, their size and complexity and the growing demands upon them are all contributing factors to systemic risk in these complex systems.

Leading Practice: Boards should engage third-party experts to do periodic systemic cyber risk reviews or audits to identify systemic risks and to verify their and management's understanding of the organization's distributed and systemic risk environment.

Systemic risk is an inherent problem resident within the concepts of complex systems science. Knowing that this issue exists for a reason helps boardrooms and leaders understand the problem and take actions that can effectively mitigate its risks — as well as learn from other domains in complex systems science.

The DDN DiRECTOR™ Framework © DDN LLC

Nowhere is systemic cyber risk more in focus than at DOMINO. DDN has pioneered the only framework for governing and identifying systemic cyber risk in complex digital systems known as DiRECTOR™. Over 600 directors, CIO's, CISO's and others from leading companies around the world have been trained and certified on it.

Systemic risk reflects the inherent risk within and between the parts of complex systems that can jeopardize the purpose of the entire system because of the failure of one part within the system. In a world of distributed risk that is inherent within modern digital business systems, a collective security view and model is required, but unfortunately largely absent in reality.

DiRECTOR is focused on changing that. DiRECTOR is part of the BLAST RADIUS-FALLOUT (BRFO) process for cybersecurity incident materiality determination. DiRECTOR is also being used by some of the world's leading companies as an integrated part of their overall enterprise risk management program.

Leading Practice: Educate directors on concepts of complex systems science and ensure their overall curriculum regularly covers updates across all three dimensions of their risk responsibilities, e.g., opportunity, cybersecurity and systemic cyber risk. Board's should ensure that management teams monitor levels of systemic risk, including third-party risk and have controls in place that are focused on establishing systemic resiliency.

Lessons From The Living Systems All Around Us

One of the most anticipated talks at DOMINO is always Bucknell University's Professor Joseph Tranquill. Professor Tranquillo always stretches the thinking of attendees with what we can learn from complex systems science.

At DOMINO 24 he brought his insights on the issues of how complex systems heal to help attendees understand the issue and ultimate goal of building resiliency within the context of the digital business system. He has translated his talk for us into the following contribution to The DOMINO Guide.

How can companies cope with a cyber threat landscape that is expanding faster than their abilities to defend against what needs to be defended? These threats are natural, i.e., inherent within the system and often artificially introduced by those wishing to harm the digital system. Perhaps the answer lies in how living systems manage similar challenges, or don't.

The goal in making these connections is to begin to view complex digital business systems as dynamic environments where the ultimate goal is not to make them 100% secure, which cannot be done, but to make them as resilient as they need to be for the purposes that they exist.

Living systems are based on four general patterns. First, they are open to their environment but functionally closed. They exist in a sweet spot of having all of the functions they need to survive inside of them (functional closure), but still allow in the most basic resources and send out "waste" (open to the environment). This boundary defines what the system can do, and cannot do.

Second, they have a metabolism. This is the sum total of all flows coming in and out as well as transformations inside the system.

Third, living systems maintain a homeostasis. That is, the ability to keep many parameters within acceptable ranges to keep all of the flows going. These first three principles are maintained dynamically, with functions being ramped up or down depending upon the circumstances and stimulus the system is experiencing.

Health of the system is directly dependent upon the system being able to maintain a boundary, regulate and maintain its metabolism and achieve homeostasis, a sort of dynamic stability and equilibrium.

Disease and death occur when these patterns cannot be sustained.

A cybersecurity incident is therefore an interruption or disruption to the homeostasis established within a complex digital business system that causes the system to destabilize.

Complex digital business systems are comprised of a lot of different types of interconnected parts. ISO 15288 describes a digital system as:

> ...a combination of interacting elements organized to achieve one or more stated purposes. The interacting elements that compose a [digital] system include hardware, software, data, humans, processes, procedures, facilities, materials, and naturally occurring entities.

Keeping such a system functioning securely is no easy task. Living systems are also constantly facing challenges, both internal (e.g. high blood pressure) and external (e.g., being in a car accident), that requires a fourth pattern – defenses along with the capacity to heal and learn.

Defenses are meant to be preventative, but some attacks will still get through, as is the case with cybersecurity risk management as no digital system can be 100% secure. A reality based upon the concepts of complex systems science.

If a threat is successful, then secondary defenses are activated where healing will repair any damage that is done by returning the system to the status quo. Learning occurs and is necessary when it is not possible to go back to the way things were and the system then creates better preventative defenses. This happens constantly in cybersecurity as security teams learn about new threats.

In this abstract form, you can think of a digital business system as a living system, maintaining a boundary that is open but functionally closed (fulfilling the purpose it was created for, but no more), a metabolism that transforms inputs of lower value (i.e., data) into an output of higher value (i.e., knowledge, insight and/or action) and a homeostasis that balances the many measures of a healthy organization.

It makes sense then for a digital business system to also have a systemic way to defend itself, heal when damaged and learn to preempt future attacks — this drives systemic resiliency.

Over billions of years living systems have evolved five sub-systems that work together to defend, heal and learn. Simple systems have a few, while more complex systems have them all. You can abstract these to build a comprehensive cybersecurity system capable of making any digital business system as resilient as it needs to be.

First is an early detection system composed of sensors and reflexes. You never know when an attack is coming. A sensor system must always be vigilant — watching all of the entrances, exits and ideally any pinch points in the system. If a sensor detects something that is creating immediate danger, most living systems have a reflex system with a very simple pre-programmed response to minimize damage as quickly as possible. An example is pulling your hand back from a hot pot, which happens without checking in with your brain.

In a digital setting, these reflexes are often automated responses to unwanted behaviors, such as repeatedly entering an incorrect account password which can result in the account being disabled or monitoring employee behaviors within the IT system such as identifying high risk e-mails and blocking them.

Suspicious network communications can also create security event alarms in surveillance systems, which may trigger automation, such as system isolation.

Second is an innate immune system that is slightly slower than a reflex but is a systemic response. A broadcast system is needed to alert the entire system, which can then roll out some pre-programmed general responses. In a living biological system this is often cleaning away toxins such as by swelling the area with fluid to dilute the damage, or rising the temperature to "cook" a virus. This system also must have an "off switch" to control against the innate immune system beginning to do its own damage.

In a digital setting, this broadcast system is a Security Operations Center (SOC) monitored by humans who initiate a triage process to quickly assess and prioritize suspect security events. Information is gathered, and cyber incident playbooks are deployed, prioritizing and allocating appropriate resources to respond according to the incident's threat, impact or potential impact. This method allows the system to manage its limited resources effectively.

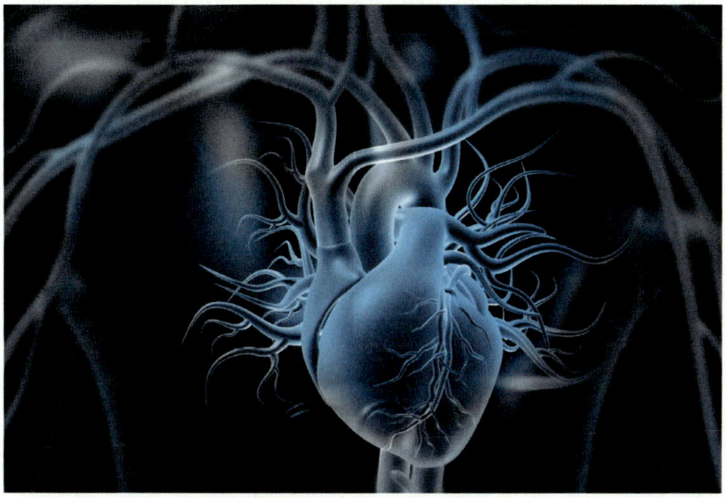

Third is a slower acquired immune system that can create a unique, targeted and more efficient response to a specific attack. The early detection and innate immune systems buy time and mitigate damage to characterize the threat.

Incident response plans experience this as more information begins to be understood about an incident from its initial discovery and the BLAST-RADIUS and FALLOUT of the incident begins to be understood. The BRFO process for cybersecurity incident materiality determination is addressed earlier in The DOMINO Guide.

If the attacker, or nature of threat, has been seen before and there is already a solution resident within the system, this can be rolled out quickly. A flu shot works this way.

If not, then a new solution must be prototyped and tested, usually through recombination of some primitive parts, in parallel, and on a small scale. When a suitable solution is found, it is then scaled to the entire system. This solution is stored so if the same (or similar) attack is encountered in the future this solution can be ramped up quickly as well. Over time the acquired immune system becomes a library of solutions to past attacks. In biological systems, this underlies the idea of immunity and vaccinations.

Fourth is a healing and learning system. When damage has been done, healing returns the system to what it was before, perhaps with some scar tissue. This is why data backups are emphasized in cybersecurity.

Sometimes during the healing process, the system needs targeted strengthening, which is the purpose of slings and casts during recovery, or hiring cybersecurity experts during incident response and recovery. Systemic resilience can also be achieved through redundancy such as having two arms, eyes, ears or kidneys. In this way, the loss of a limb allows the body to continue to perform its tasks in a less-optimal, but still functional manner.

In digital systems, resilience means designing systems to absorb cyber-attacks while continuing to function, even if in a less-optimal manner. For example, some digital systems use multi-cloud capabilities to provide geographic redundancy and avoid single-point-of-failure issues with single cloud providers.

Learning is a move to a new way of staying alive, of strengthening and extending the boundary, metabolism or homeostasis to build resiliency. Healing and learning often happen together — repairing damage but in a way that leaves the system better prepared the next time.

Fifth is a thinking system that recognizes threats ahead of time and takes preventative or anticipatory actions. Threat hunting in cybersecurity deploys this concept to create proactive resiliency.

Recognizing the signs of a potential mugger on the street, and avoiding the threat by walking into a store for example. The more a system knows about the world the better the thinking system can be — animals don't know about germs so they don't wash up, but humans do. But instinctively, they have learned to clean themselves for similar reasons.

Like the immune system, the thinking system has fast and slow parts. The fast part are emotions. Pre-programmed heuristics that have physiological impacts (e.g., heart rate, eye dilation) and guide our behavior. They evolved to create sensory instinct so that we know to "pay attention, something important is happening".

The slow part is our rational mind that can plan, contemplate scenarios, learn from failure and direct intentional practice. It can even (in more advanced organisms) change the environment to make it more suitable or safer.

Companies can view their entire cybersecurity system through these same five sub-systems to understand the policies, processes and practices they need to have in place to create systemic resiliency in the digital business systems that power their business. The ways living systems defend, heal and learn may be helpful in making cybersecurity concerns and strategies more understandable and relatable to non-technical employees, the C-suite and even within the boardroom.

Considering a digital business system as a living system can also serve as a compostable architecture, one that can adapt to the risks and promises of new technologies — from AI to quantum computing to self-healing materials.

The health of a system is not only about being free from disease. Healthy systems have the bandwidth to learn and innovate in the face of changes to their environment. The same systems mindset and functions needed to defend, heal and learn can also be turned to innovation. The sensors that are always looking for threats can also recognize untapped opportunities. A system that learns from failures can also learn from successes. Understanding risk in this context can lead to new solutions to eliminate unacceptable levels of risk to the system. The run flat tire was invented for this reason. A broadcast system that sends out alerts for threats can also call attention to opportunities.

A system that can test and scale a solution to a problem can also scale a value-creating opportunity.

So learn from the systems around us and the complex system that we are, to understand how to build systemic resiliency from a digital and cybersecurity perspective — that should be the goal of every board, CISO and CEO, as it is the goal of all living systems.

Next Steps: Strengthening The Board To Optimize The Digital Business System

The last word at DOMINO 24 was heard from some of the boardroom leaders who are driving boardroom transformation on digital, cybersecurity and systemic cyber risk governance. They are creating high-performing boards that are governing the challenges of complex digital business systems and adding value to their companies as a result.

Solving this problem is not just possible, it is a necessity, and it is already being solved.

At DOMINO 24 a group of corporate directors who are at the forefront of solving these challenges shared their insights. **Ed Magee, Director at WD-40 (NASDAQ: WDFC) shared that in his experience moving this conversation with corporate directors out of just cyber risk to include business opportunities is a very useful tactic for engaging directors on the details.**

Ed also said, "Leadership scar tissue on these issues is invaluable. Learn from other's mistakes." He also believes that directorship should be approached as a profession, with directors taking responsibility for continual learning on the many issues facing their companies, including digital, cybersecurity and systemic cyber risk governance.

Daphne Jones, DDN.QTE and former CIO at GE Healthcare Medical Services and Director at Barnes Group (NYSE: B) and AMN Healthcare (NYSE: AMN), shared that "Digital resilience is not just about preventing and detecting disruptions but thriving in spite of them. Boards that lead ask great questions about breach recovery planning AND they must be able to understand the implications of the answers."

Daphne is someone who has been leading with a deep understanding of these issues to help her boards become more effective, She broke her efforts down into the three stages of development on these issues. **First, she consciously took her boards on a journey of getting them interested. Engagement followed and then commitment set in once there was understanding.** Once her fellow directors got engaged and started learning about these issues in the right context, she said the commitment started to come naturally.

How could it not? What CEO or investor would not want a corporate board with the deep capabilities to oversee these complex issues?

While coercing change in the boardroom through regulatory mandate has just started, self-regulation is the leading practice in transforming the boardroom in digital, cybersecurity and systemic risk governance.

Kris Lovejoy, a corporate director at Dominion Energy (NYSE:D) commented, "Making this just a cybersecurity issue in the boardroom is too narrow of a view. Talk about business value, and the board will listen all day long."

Corporate boards can and should transform their approach to digital, cybersecurity and systemic risk governance in the short-term through self-regulatory reform. They do not need to — nor should they — wait for regulators or hackers to do it for them, or to them. It is not expensive, or burdensome to adopt the many leading processes, policies and practices included in The DOMINO Guide or those created by the standards bodies and leading practitioners.

Most importantly when the boardroom is a high- performing part of the organization's system of digital value creation and protection, evidence shows that this creates significant amounts of business value.

DDN and DOMINO 24 brought together the leaders who are defining the leading edge of these issues to establish a blueprint that any boardroom can follow and apply. DOMINO 24 taught attendees how to learn from, apply and advance the three layers of this puzzle including:

1. Standards.
2. The board's system of governance.
3. The proper scope of digital risk oversight.

Legacy views and status-quo approaches are not solving these problems. The boardroom and executive leaders at DDN and DOMINO 24 are. They are collectively stepping up to this leadership moment to shape thoughtful, transparent, and practical processes that will shape and secure the path into the digital future for their companies.

Building the practice and profession of digital, cybersecurity and systemic risk oversight is our goal at DDN, a goal we're well on the path to achieving.

In the words of Daphne Jones to those at DOMINO 24, "We are the enlightened. We have to move with intentionality, to meet our boards where they are, but to have a plan to move them to where they need to be on these issues."

DDN CEO Bob Zukis wrapped up DOMINO 24 with this call to action for both sides of the boardroom table:

Use The SEC's Gift — CISO's have been gifted a leadership moment from the SEC. Use cybersecurity incident materiality tabletops to raise your profile, lead the analysis and be engaged in the disclosure process with leadership and the board. Regulatory accountability is beginning for corporate directors and boardrooms. Hold yourself to a higher standard to avoid leering hackers be your regulator.

Be An Educator — Recognize the board as a part of the system — as a control to be strengthened on digital value creation and protection. Let leading practices and established standards do the educating for you about the value of being a high-performing board on these issues. As a director, know that effective digital and cybersecurity oversight drives business value.

Don't undervalue the importance of your role as a CISO and know how much revenue and profitability you wake up to protect every day. As a CIO or CTO recognize that they company's value propositon goes through your office, and that you have the board's full focus.

Be The Systemic Cyber Risk Champion — The board and C-suite is now awake to systemic cyber risk given the CrowdStrike incident. But systemic cyber risk, and systems thinking is likely a new concept for the board and C-suite. Step up to this dimension of risk given its prevalence throughout the digital business system to engage and educate business leadership and demonstrate your value as an enterprise risk management leader or value creator.

At DDN we invite you to join us on this journey—your shareholders and other stakeholders will thank you. We are pleased to share with you these learnings from DOMINO 24 in The DOMINO Guide: *The Definitive Boardroom Guide on Digital, Cybersecurity And Systemic Risk. Governance.*

Footnotes

[i] Weill, Peter et al., Companies With A Digitally Savvy Board Perform Better. Research Briefing Volume XIX, Number 1, January 2019. MIT CISR.

[ii] Lee, Jennifer. Pillay, Shoba. Riely, Charles D. Takeaways from SEC v. SolarWinds Motion to Dismiss Hearing. June 9, 2020. Harvard Law School Forum on Corporate Governance. https://corpgov.law.harvard.edu/2024/06/09/takeaways-from-sec-v-solarwinds-motion-to-dismiss-hearing/#:~:text=The%20 SEC's%20high%2Dprofile%20litigation,on%20Defendants'%20motion%20to%20dismiss.

[iii] https://www.sec.gov/newsroom/press-releases/2023-227

[iv] ISO/IEC 27014:2020 (E) Information security, cybersecurity and privacy protection —Governance of information security 2nd Edition (2020).

[v] National Institute of Standards and Technology U.S. Department of Commerce, The NIST Cybersecurity Framework (CSF) 2.0, February 26, 2024

[vi] Isidore, Chris. Delta CEO lashes out at CrowdStrike: This cost us $500 million and they offered us nothing. CNN. July 31, 2024. https://www.cnn.com/2024/07/31/business/delta-meltdown-costs/index.html

[vii] Cybersecurity Risk Management, Strategy, Governance, and Incident Disclosure, Release No. 33-11216, July 26, 2023 [2023-16194 (August 4, 2023)] (Final Rule)

[viii] Digital Directors Network. SEC Leadership on Cybersecurity Incident Materiality, May 29, 2024. https://www.linkedin.com/events/ddntalkswithsecleadershiponcybe7199167717268967426/theater/

[ix] European Parliament. EU AI Act: first regulation on artificial intelligence. www.europarl.europa.eu, June, 18, 2024. https://www.europarl.europa.eu/topics/en/article/20230601STO93804/eu-ai-act-first-regulation-on-artificial-intelligence

[x] Ibid.

[xi] "DDN Calls for a Technology & Cybersecurity Committee on US Public Company Boards." Business Wire. DDN LLC, September 5, 2018. https://www.businesswire.com/news/home/20180905005332/en/DDN-Calls-for-a-Technology-Cybersecurity-Committee-on-US-Public-Company-Boards.

[xii] Chen, Kevin D., and Andy Wu. "The Structure of Board Committees." Harvard Business School Working Paper 17-032, (2016).

[xiii] FedEx Corporation. Cyber and Technology Oversight Committee Charter. March 11, 2024. https://s21.q4cdn.com/665674268/files/doc_downloads/committee_charters/2024/03/Cyber-and-Technology-Oversight-Committee-Charter.pdf

[xiv] CONSTRUCTION INDUSTRY LABORERS PENSION FUND, CENTRAL LABORERS' PENSION FUND, LAWRENCE MILES, and BRIAN SEAVITT, derivatively on behalf of SOLARWINDS CORPORATION, Plaintiffs, vs. MIKE BINGLE, WILLIAM BOCK, SETH BORO, PAUL J. CORMIER, KENNETH Y. HAO, MICHAEL HOFFMANN, DENNIS HOWARD, CATHERINE R. KINNEY, JAMES LINES, EASWARAN SUNDARAM, KEVIN B. THOMPSON, JASON WHITE, MICHAEL WIDMANN, (Court of Chancery of the State of Delaware November 4, 2021).

APPENDIX — DDN Key Policy Positions

DIRECTOR SKILLS: Digital and Cybersecurity Expertise

DDN POLICY POSITION: Private and public company board composition should include directors with the expertise and experience required to effectively govern all domains of the complex digital business system and its role in the organization's value proposition. Every public company board should have at least one director who can be named as a director cybersecurity expert.

RATIONALE: Director expertise specific to complex digital business systems is proven to drive enhanced business outcomes and reduce cybersecurity risk. Strengthening the boardroom as a leadership control in the digital and cybersecurity system also strengthens the entire digital and cybersecurity system.

Boardroom qualified technology experts (QTE's) should have expertise gained from relevant applied work experience in one or more domains of a complex digital business system and/or education and knowledge demonstrated by professional and accredited verification. Demonstrated and applied aptitude is the relevant determinant of expertise as opposed to job-titles.

Demonstrated domains of relevant expertise and experience should include data, information architecture, risk communications, emerging technology, cybersecurity, third-party and systemic risk, IT operations, relevant regulatory experience. All Directors should receive annual training on digital, cyber-security and systemic risk.

THE BOARD's STRUCTURE: Technology and Cybersecurity Committee

DDN POLICY POSITION: Private and public corporate boards should consider as a leading practice organizing digital, cybersecurity and systemic cyber risk responsibilities within a Technology and Cybersecurity Committee (TCC). Scope of the TCC should govern the value creation and protection issues related to upside opportunity risks, downside cybersecurity risks and systemic risk related to the role of the complex digital business system in the organization's overall value proposition.

RATIONALE: A Technology and Cybersecurity Committee brings task efficiency, focus and greater accountability on digital issues for investors, the full board and management team as a strong internal and external leadership control. A comprehensive charter is vital to ensure that the board is focused on the breadth and depth of the entire digital system from it's implementation to operation.

Where a Risk Committee already exists consideration should be given to expanding this charter and committee composition to include the oversight of opportunity, cybersecurity and systemic risk related to the digital business system.

DDN recommends against the practice of cyber risk being overseen by an audit committee given marginalization of these complex issues to the broader audit committee mandate/agenda and audit committee director member competency misalignment. DDN recommends that there be multi-committee director representation from the audit committee, or expressly stated cross committee coordination with the committee where enterprise risk responsibility resides to reduce information asymmetries.

THE BOARD's SCOPE OF OVERSIGHT: Opportunity, Cybersecurity and Systemic Risk

DDN POLICY POSITION: The boardroom is a key control in the complex digital business system for both business value creation and protection. The scope of risk oversight in corporate governance needs to span three aspects of risk related to complex digital business systems including opportunity risk, cybersecurity risk and systemic risk.

RATIONALE: Based upon empirical evidence, significant business value benefits are demonstrated when directorship is a high-performing part of the organization's digital value creation and protection systems.

We support NIST CSF 2.0, ISO 27014 and 38500 as baseline policy and practice models for establishing a scope of boardroom oversight in opportunity risk and cybersecurity risk. We advocate for use of the DDN DiRECTOR™ framework as the leading practice in systemic cyber risk identification and governance.

Board policy should also extend to monitoring and developing leading policies and practices that extend beyond these models to cover all applicable policy and process areas including:

- Business value and digital strategy alignment
- Enterprise information architecture effectiveness and resiliency
- Business continuity and disaster recovery
- Risk communications
- Cyberthreat intelligence
- Data and information lifecycle management, including privacy
- Device management
- Regulatory compliance and government relations in cybersecurity and data privacy
- IT and cyber operations effectiveness
- Project prioritization, implementation, and portfolio management
- Digital skills and capability management
- Digital hardware/software lifecycle management
- New and emerging technology opportunity and risk
- Social media monitoring and management
- Third-party and systemic cyber risk management
- Cyber risk economic impact analysis and determination

Effective digital, cybersecurity and systemic risk governance is a complex system in and of itself that needs the right skills, structure, and scope of risk oversight to be a high-performing part of corporate governance. Leading practices continue to emerge in all three areas and should be monitored to optimize the role of the board in strengthening the performance and resiliency of the organization's complex digital business system in the creation and protection of business value.

APPENDIX — DDN SEC Cybersecurity Disclosure Cheat Sheet

REQUIRED FORM 8-K CURRENT REPORT DISCLOSURES: MATERIAL CYBERSECURITY INCIDENTS (1)
Upon determining that a cybersecurity incident (unauthorized or accidental occurrence) is material to a reasonable investor, SEC registrants are required to disclose the following information within 4-business days in Section 1.05 of Form 8-K as long as it is not an issue of national security or public safety as determined by the U.S. Department of Justice (2). Cybersecurity incidents not material or yet deemed material, can also continue to be disclosed if important in Section 8.01 of Form 8-K.
☐ Describe the material aspect of the nature, scope, and timing of the incident (3).
☐ Describe the material impact or reasonably likely material impact on the registrant, including (non-exclusively) financial condition and results of operations along with qualitative factors such as harm to reputation, customer of vendor relationships, competitiveness, litigation, U.S. and non-U.S. regulatory investigations, etc.
☐ Disclose updated [material] incident information within four business days, after the registrant, without unreasonable delay, determines such information or within four business days after such information becomes available in a Form 8-K amendment.
(1) Cybersecurity incident means an unauthorized occurrence, or a series of related unauthorized occurrences, on or conducted through a registrant's information systems that jeopardizes the confidentiality, integrity, or availability of a registrant's information systems or any information residing therein. **Accidental occurrences are also an unauthorized occurrence**. Information system means electronic information systems owned or used by registrant, including physical or virtual infrastructure controlled by such information resources, or components thereof, organized for the collection, processing, maintenance, use, sharing, dissemination, or disposition of the registrant's information to maintain or support the registrant's operations. The materiality determination is be made without unreasonable delay.
(2) If at any time incident is deemed to be in the interest of national security or public safety, the FBI is to be contacted to begin the process for disclosure delay consideration. Otherwise, disclosure is due within 4-business days of determining it is material to a reasonable investor. https://www.fbi.gov/investigate/cyber/fbi-guidance-to-victims-of-cyber-incidents-on-sec-reporting-requirements
(3) Disclosure does not need to include "specific or technical information about its [registrants] planned response to the incident or its cybersecurity systems, related networks and devices, or potential system vulnerabilities in such details as would impede the registrant's response or remediation of the incident."

REQUIRED FORM 10-K ANNUAL REPORT DISCLOSURES: RISK MANAGEMENT, STRATEGY AND GOVERNANCE OF CYBERSECURITY RISKS
RISK MANAGEMENT AND STRATEGY
Registrants are now required to provide more consistent and informative disclosure regarding their cybersecurity risk management and strategy in their annual reports, Form 10-K, by:
☐ Describing the registrant's processes, if any, for assessing, identifying, and managing material risks from cybersecurity threats in sufficient detail for a reasonable investor to understand those processes (4). The enumerated elements, non-exclusively (5), that a registrant should specifically address within this disclosure include:

☐ Whether the above processes have been integrated into the registrant's overall risk management processes;
☐ Whether the registrant engages assessors, consultants, auditors, or other third parties in connection with any such processes;
☐ Whether the registrant has processes to oversee and identify **material** risks from cybersecurity threats associated with its use of any third-party service provider and other information necessary for a reasonable investor to understand their cybersecurity processes.
☐ Describing whether any risks from cybersecurity threats, including as a result of any previous cybersecurity incidents, have materially affected or are reasonably likely to materially affect the registrant, including its business strategy, results of operations, or financial condition and if so, how.
MANAGEMENT'S ROLE AND CORPORATE GOVERNANCE
☐ Describing the board's oversight of risks from cybersecurity threats.
☐ Including, if applicable, identifying and describing if any board committee, or subcommittee is responsible for such oversight
☐ Describing the processes by which the board or such committee is informed about such risks
☐ Describing management's role in assessing and managing the registrant's material risks from cybersecurity threats including describing (non-exclusively):
☐ Whether and which management position or committees are responsible for assessing and managing such risks, and
☐ The relevant expertise of such person or members in such detail as necessary to fully describe the nature of the expertise (6).
☐ The processes by which such persons or committees are informed about and monitor the prevention, detection, mitigation, and remediation of cybersecurity incidents.
☐ Whether such persons or committees report information about such risks to the board of directors or a committee or subcommittee if the board.
(4) Processes are not policies and procedures. This context is intended to allow registrants to avoid operational details that may advantage threat actors and exacerbate risk. The SEC expects this disclosure to be sufficient enough to allow investors to ascertain a registrant's cybersecurity practices, such as whether they have a risk assessment program in place, with sufficient details for investors to understand the registrant's cybersecurity risk profile, i.e. risk appetite and tolerances.
(5) Registrants should additionally disclose whatever information is necessary based on their facts and circumstances for a reasonable investor to understand their cybersecurity processes.
(6) Prior work experience in cybersecurity; any relevant degrees or certifications; any knowledge, skills, or other background in cybersecurity.

APPENDIX — Who Is FOR And Who Is AGAINST Director Cybersecurity Expertise In The Boardroom

Who Is FOR Director Cyber Expertise Disclosure, And Why?

We know the hackers are AGAINST competent cybersecurity leadership in the boardroom, as are some others. But who was **FOR** the proposed U.S. SEC boardroom director cybersecurity expertise disclosure rule, and why?

Given that a disclosure requirement of director cybersecurity expertise would have had the causal effect of driving SEC registrants to add directors to the board with true cybersecurity expertise, analyzing this issue is a close proxy for the more substantive boardroom reform issue of who supports the need for boards to have resident director cybersecurity expertise.

This analysis lists and reflects the different statements and arguments from those who were FOR vs. AGAINST the boardroom cybersecurity disclosure requirement based upon their comment letters submitted to the SEC in 2022. The cybersecurity disclosure rules were finalized in 2023 and went into effect in December 2023 for most issuers — without the director cybersecurity expertise disclosure requirement which was dropped in the final rules.

A core DDN policy is that strengthening the board as a control in the cybersecurity system requires directors with cybersecurity expertise. DDN has also made a call for every U.S. public company board to add a director with cybersecurity expertise. Other groups and organizations are also supportive of the proposed director cyber-security expertise disclosure rule, but who were they and what were their views?

The comments and arguments of those FOR the disclosure rule reflect empirical evidence through actual experience with the issue on both sides of the boardroom table. The FOR's were also a very different group than those who were in the AGAINST camp.

The following information is compiled from the SEC comment letters submitted by the organization in response to the proposed SEC cybersecurity disclosure rules. They are public information. It may also include other public and documented comments as indicated.

We analyzed the entirety of each submission and identified what we felt were the most specific comments related to the proposed disclosure rule which would have required director cybersecurity expertise disclosure. The initial SEC proposal, which these opinions and comments were solicited on, was stated as follows:

> We propose to amend Item 407 of Regulation S-K by adding paragraph (j) to require disclosure about the cybersecurity expertise of members of the board of directors of the registrant, if any. If any member of the board has cybersecurity expertise, the registrant would have to disclose the name(s) of any such director(s), and provide such detail as necessary to fully describe the nature of the expertise. — SEC Proposed Rule

As indicated, this was merely a disclosure requirement. It was not a requirement to add cyber expertise to the board, just a step towards bringing transparency for investors as to whether this expertise was present.

The FOR cohort reflects industry executives and practitioners, institutional investors, cybersecurity software providers, academia and government policy makers. **We summarize the FOR group as those who know the details of cybersecurity and those who are focused on protecting investors and other stakeholders.**

Arguments presented by those who were AGAINST the proposed cyber expertise disclosure rule were much different than those in the FOR camp. The FOR camp also included empirical evidence for their views, while the AGAINST camp used a collection of fallacies, and a coordinated propaganda campaign to advocate against this issue.

CYBERSECURITY AND ACCOUNTING INDUSTRY ASSOCIATIONS

WHO IS **FOR** BOARDROOM DIRECTOR CYBER EXPERTISE DISCLOSURE?
Cybersecurity and Accounting Industry Associations – Digital Directors Network, ISC2, SANS Institute, BSA Software Alliance, The Cipher Brief, AICPA
Institutional Investors and Investors Associations - CalPERS, CII, PRI, NASAA
Cybersecurity Software Leaders - Crowdstrike, Tenable
Government Policy Leaders – Seven bi-partisan U.S. Senators
Academia - Virginia Tech, Columbia University, Texas A&M School of Law, Prairie View A&M University, Michigan State, The University of Arizona.

DIGITAL DIRECTORS NETWORK

The premier boardroom network of over 1,600 corporate directions, CIOs, CISOs and other technology executives working to advance the practice and profession of digital, cybersecurity and systemic risk governance.

What We Said: The item E amendment [boardroom cybersecurity expertise disclosure] is singularly the highest impact, lowest effort proposal being suggested that will materially lower cyber risk exposure for America's public companies. Strengthening America's boardrooms as a critical control point in cybersecurity is an issue of national security. This issue is also core to the competitiveness of American businesses and one squarely in the interest of investors and consumers. All other proposals in S7-09-22 will be strengthened through this foundational, common-sense, overdue, and easily implementable boardroom reform.[1]

ISC2

ISC2 is an international not-for-profit membership association of more than 170,000 certified cyber security professionals, focused on inspiring and delivering a safer and more secure cyber world.

What They Said: ISC2 supports proposed item 407(j) [disclosure of board member cybersecurity expertise] and forms the opinion that investors would find it useful. The proposed item does not impose any additional obligations on the directors. Strategically, proposed item 407(j) could prove beneficial as it demonstrates commitment to cybersecurity which would elevate confidence in the organization with the public and other entities.

ISC2 supports the requirement that corporate officeholders, including directors, demonstrate a level of expertise in cybersecurity commensurate with the organizational risk that cybersecurity presents in the digital era. Information risk represents a set of risks that holds the same level of importance in terms of potential impact as financial risk, human risk and other major organizational risks. As such, information security requires a similar level of scrutiny and controls at a board level.[2]

SANS INSTITUTE

The SANS Institute is a private U.S. for-profit company founded in 1989 that specializes in information security, cybersecurity training, and certification. Topics available for training include cyber and network defenses, penetration testing, incident response, digital forensics, and auditing.

What They Said: Requiring disclosure of cybersecurity expertise would be a net positive.

2018/2019 SANS Project found similar points about Board Members and cybersecurity

- Ability to recognize when strategies were or were not likely to lead to changes needed to reduce overall cybersecurity risk.
- Going beyond compliance or liability focus to broader security/safety strategy evaluation[3].

Footnotes

1. Zukis, Bob. Re: FILE NUMBER S7-09-22 — COMMENTS ON CYBERSECURITY RISK MANAGEMENT, STRATEGY, GOVERNANCE, AND INCIDENT DISCLOSURE. Digital Directors Network. May 8, 2022. https://www.sec.gov/comments/s7-09-22/s70922-20128269-290188.pdf
2. SEC PROPOSED RULE: CYBERSECURITY RISK MANAGEMENT, STRATEGY, GOVERNANCE, AND INCIDENT DISCLOSURE. ISC2. May 9, 2022. https://www.sec.gov/comments/s7-09-22/s70922-20128411-291326.pdf
3. Greber-Raines, Ian. File No. S7-09-22. SANS Institute. June 7, 2022. https://www.sec.gov/comments/s7-09-22/s70922-20130968-300269.pdf SEC PROPOSED RULE: CYBERSECURITY RISK MANAGEMENT, STRATEGY, GOVERNANCE, AND INCIDENT DISCLOSURE. ISC2. May 9, 2022. https://www.sec.gov/comments/s7-09-22/s70922-20128411-291326.pdf

BSA | THE SOFTWARE ALLIANCE

BSA is the leading advocate for the global enterprise software industry before governments and in the international marketplace. Its members are among the world's most innovative companies, providing the products and services that power governments and businesses. BSA's members offer software that generates efficiencies and promote trust and security, including cloud computing, customer relationship management, human resources management, and identity and access management products and services.

What They Said: Maintaining appropriate technical and administrative controls as part of a comprehensive, risk-based cybersecurity risk management program, with effective oversight, is critical to managing cybersecurity risk. Additional public transparency on these key points will both encourage sound practices and help investors make informed judgments about the effectiveness of a registrant's cybersecurity risk management program. BSA accordingly supports disclosure of limited relevant information about a registrant's cybersecurity policies and procedures and its board of directors' cybersecurity expertise.[4]

THE CIPHER BRIEF - CYBER INITIATIVES GROUP

The Cyber Initiatives Group, a committee formed and sponsored by The Cipher Brief, is a private media organization that engages with the private sector in the United States to promote awareness of cybersecurity and national security matters. Many of its members currently have direct involvement in cyber matters in the private sector and have significant experience in both policy and operational aspect of cybersecurity; many of its members have served at the highest levels of the U.S. armed forces or intelligence community, while others have leading roles at the nation's most significant cybersecurity firms and technology providers. Writing in individual capacities.

What They Said: We believe that the goals of requiring current reporting about material cybersecurity incidents, as well as periodic disclosures regarding (1) a registrant's policies and procedures to identify and manage cybersecurity risks, (2) management's role in implementing cybersecurity policies and procedures and (3) the board of directors' cybersecurity expertise and its oversight of cybersecurity risk, are appropriate and are likely to enhance the cybersecurity posture of registrants.

Consequently, improved cybersecurity within public companies translates directly into a national economy that is more cybersecure and cyber-resilient. It stands to reason that requiring additional reporting about material cyber incidents will better inform investors, the public generally and governmental agencies, and increased disclosure about cyber policies and board experience will encourage public companies (and by extension, private companies, at least to some degree) to meet if not exceed market expectations in those areas).[5]

AICPA

The American Institute of Certified Public Accountants is the national professional organization of Certified Public Accountants in the United States, with more than 428,000 members in 130 countries

What They Said: The AICPA agrees that investors and other stakeholders may benefit from disclosures related to board members' cybersecurity expertise. Understanding such matters enables investors to consider the importance a registrant places in effectively managing its cybersecurity risks. As indicated in Appendix A, the AICPA's Description Criteria for Management's Description of the Entity's

Footnotes

4. Young, Henry. BSA The Software Alliance. May 9, 2022. https://www.sec.gov/comments/s7-09-22/s70922-20128387-291290.pdf
5. Bissell, Kelly. et al. File Number S7-09-22 – Comments on Proposed Rule. May 9, 2022. https://www.sec.gov/comments/s7-09-22/s70922-20128383-291286.pdf
6. Coffey Susan. Re: Request for Comment on Proposed Rule, Cybersecurity Risk Management, Strategy, Governance, and Incident Disclosure [Release Nos. 33-11038; 34-94382; IC-34529; File No. S7-09-22]. AICPA. May 6, 2022. https://www.sec.gov/comments/s7-09-22/s70922-20128331-291060.pdf

Cybersecurity Risk Management Program also cover disclosure of "the process for board oversight of the entity's cybersecurity risk management program."[6]

INSTITUTIONAL INVESTORS AND INVESTORS ASSOCIATIONS

CalPERS

The largest public defined benefit pension fund in the United States, CalPERS manages approximately $470 billion in global assets on behalf of more than 2 million members.

What They Said: Beginning with the Cybersecurity Disclosure Act of 2017 (S. 536 by Senator Jack Reed (D-RI)), we have consistently supported legislative efforts to require publicly traded companies to disclose in their annual reports or annual proxy statements, whether any member of their governing body, such as a board of directors, has expertise or experience in cybersecurity issues.[7]

COUNCIL OF INSTITUTIONAL INVESTORS

CII is a nonprofit, nonpartisan association of United States (U.S.) public, corporate and union employee benefit funds, other employee benefit plans, state and local entities charged with investing public assets, and foundations and endowments with combined assets under management of approximately $4 trillion. It's member funds include major long-term share-owners with a duty to protect the retirement savings of millions of workers and their families, including public pension funds with more than 15 million participants — true "Main Street" investors through their pension funds. Our associate members include non-U.S. asset owners with about $4 trillion in assets, and a range of asset managers with more than $40 trillion in assets under management.

What They Said: Directors have the authority, capacity and responsibility to make pivotal contributions in this area [cybersecurity] by ensuring adequate resources and management expertise are allocated to robust cyber risk management policies and practices, and ensuring disclosure fairly and accurately portrays material cyber risks and incidents.

To achieve these objectives, directors need to:

- Understand management's cybersecurity strategy;
- Learn where cybersecurity weaknesses lie and;
- Support informed, reasonable investment in the protection of critical data and assets.

We are pleased to see that the Proposed Rules address the role of the board in cybersecurity risk management and strategy in a thorough manner, including disclosure of whether any board member has expertise or experience in cybersecurity.

We believe that annual disclosure of cyber expertise among board members, if any, in the annual report and proxy would be helpful to investors, especially in voting decisions.[8]

PRINCIPLES FOR RESPONSIBLE INVESTMENT (PRI)

The Principles for Responsible Investment (PRI) is the world's leading initiative on responsible investment. The PRI is a not-for-profit company with over 4,800 signatories (pension funds, insurers, investment managers and service providers) to the PRI's six principles with approximately US $121 trillion in assets under management.

What They Said: The PRI supports the inclusion of the following disclosure requirements:

- Disclosure of the cybersecurity expertise of members of the board of directors.

The PRI welcomes mandatory disclosure of board expertise on cybersecurity. In general, the PRI believes that enhanced expertise at the board level on sustainability matters is crucial to companies' sustainability efforts. Board cybersecurity expertise serves as a useful starting point for investors to assess a company's approach to cybersecurity, which is one part of sustainable operations.[9]

Footnotes

7. Frost, Marcie. Subject: Cybersecurity Risk Management, Strategy, Governance, and Incident Disclosure (File No. S7-09-22). Calpers. May 9, 2022. https://www.sec.gov/comments/s7-09-22/s70922-20128343-291102.pdf

8. Stewart, Tracy. File No S7-09-22. Council of Institutional Investors. May 9, 2022. https://www.sec.gov/comments/s7-09-22/s70922-20128381-291284.pdf

9. Kerschke, Karen. Vaz Boni, Betina. CONSULTATION RESPONSE US SECURITIES AND EXCHANGE COMMISSION: CYBERSECURITY RISK MANAGEMENT, STRATEGY, GOVERNANCE, AND INCIDENT DISCLOSURE; RIN 3235-AM89. PRI. April 2022. https://www.sec.gov/comments/s7-09-22/s70922-20125334-284791.pdf

NORTH AMERICAN SECURITIES ADMINISTRATORS ASSOCIATION

The North American Securities Administrators Association, founded in Kansas in 1919, is the oldest international investor protection organization. NASAA is an association of state securities administrators who are charged with the responsibility to protect consumers who purchase securities or investment advice.

What They Said: Second, the Proposal would require companies to disclose whether any member of the board has cybersecurity expertise and, if so, the name(s) of any such director(s) and such detail as necessary to fully describe the nature of the expertise. NASAA supports the Risk Management Disclosures, and we encourage their adoption, with certain revisions.[10]

DDN Note: No suggested revisions were offered for director cybersecurity expertise disclosure.

CYBERSECURITY INDUSTRY LEADERS

CROWDSTRIKE

A leading international, US-headquartered, cloud-native cybersecurity provider that defends globally distributed enterprises from globally distributed threats. CrowdStrike offers insights informed by multiple practice areas: cyber threat intelligence; proactive threat hunting, incident response and managed security services; and an AI-powered software-as-a-service cybersecurity platform and marketplace.

What They Said: Finally, we agree with the SEC's proposal to amend Item 407 of Regulation S-K to require disclosure about the extent to which registrants' board of directors include members with cybersecurity expertise. Based on our experience working with numerous listed companies, Board-level cyber expertise addresses three common security program failure modes:

1. Atrophy through inattention. Cybersecurity is a process that requires ongoing attention. Concerted interest and oversight from the highest levels of organizations' leadership ensure that security programs remain robust and continue to meet constantly-evolving threats and risks.
2. Compliance but ineffective security. Companies could design programs that technically meet existing or proposed compliance requirements, but do not in practice effectively address threats or mitigate risks. The focus of boards of directors on outcomes can proactively address weak or eroding performance.
3. Misaligned business conditions. Some entities lack the cybersecurity maturity to run effective security programs internally. Increasingly, such entities should rely upon managed service providers to achieve the level of security appropriate for listed companies. Organizational transformations along these lines often involve a cross section of departments or teams (e.g., personnel, finance, security, human resources) and can be most expeditiously resolved at the leadership-level.[11]

TENABLE

Approximately 40,000 organizations around the globe rely on Tenable to understand and reduce cyber risk. As the creator of Nessus®, Tenable extended its expertise in vulnerabilities to deliver the world's first platform to see and secure any digital asset on any computing platform. Tenable customers include approximately 60 percent of the Fortune 500, approximately 40 percent of the Global 2000, and large government agencies.

What They Said: Tenable commends the SEC for its initiative and ongoing efforts to enhance and standardize disclosures regarding cybersecurity, and strongly supports the SEC's proposed rule that will require public companies to disclose their policies and procedures for identifying and managing cybersecurity risks, management's role in implementing cybersecurity policies and procedures, and the board of directors' cybersecurity expertise. We believe it is important that investors know whether a registrant's board of directors has an oversight role regarding cybersecurity governance, including its oversight role regarding cybersecurity risks.[12]

Footnotes

10. Senter Lubin, Melanie. File No. S7-09-22: Cybersecurity Risk Management, Strategy, Governance, and Incident Disclosure. NASAA. May 9, 2022. https://www.sec.gov/comments/s7-09-22/s70922-20128348-291109.pdf
11. Bagley, Drew. Sheldon, Robert. Securities and Exchange Commission (SEC): Cybersecurity Risk Management, Strategy, Governance, and Incident Disclosure. Crowdstrike. May 9, 2022. https://www.sec.gov/comments/s7-09-22/s70922-20154257-322480.pdf
12. Yoran, Amit. Security and Exchange Commission proposed rule regarding Cybersecurity Risk Management, Strategy, Governance, and Incident Disclosure, File S7-09-22. Tenable. May 9, 2022. https://www.sec.gov/comments/s7-09-22/s70922-20128338-291095.pdf

GOVERNMENT POLICY LEADERS

UNITED STATES SENATE

A bipartisan group of seven U.S. Senators.

What They Said: One effective regulatory approach would be asking public companies to disclose whether a cybersecurity expert is on the board of directors, and if not, why not. We have sponsored bipartisan legislation called the Cybersecurity Disclosure Act to require companies to provide this disclosure to investors. The bill does not tell companies how to deal with cybersecurity threats. How a company chooses to address cybersecurity risks would remain its own decision. Boards of directors would be encouraged to develop approaches that address their own needs. The goal is to encourage directors to play a more effective role in cybersecurity risk oversight.[13]

ACADEMIA

VIRGINIA TECH PAMPLIN SCHOOL OF BUSINESS

US News and World Report: #47 Best National Universities and #3 Master of Information Technology Cyberdegrees.org: #1 Master of Information Technology, Cybersecurity

What They Said: Results of a research project titled: *Inexpert Supervision: Field Evidence on Boards' Oversight of Cybersecurity*

Our interviews suggest that the level of individual directors' cybersecurity expertise is an important determinant of the effectiveness of board oversight. Consistent with self-efficacy theory, cybersecurity expertise influences the attention given by boards to cybersecurity issues; when boards lack directors with cybersecurity expertise, they are less likely to substantively address cybersecurity during board meetings.

Participants described that, in the absence of expert directors, boards may adopt a compliance attitude to "check a box" by receiving cybersecurity reports without significantly engaging with cybersecurity executives. In addition, our interviewees shared that cybersecurity expertise affects the quantity, quality, and depth of the questions that directors ask cybersecurity executives as well as their ability to meaningfully respond to answers received. We find that when inexpert boards do engage with cybersecurity executives, they often ask superficial questions (such as "How are we doing?") or reactive questions (such as "Could that happen to us?") in response to a cyber incident in the media.

The Virginia Tech interviews suggest that directors' cybersecurity expertise is an important determinant of oversight effectiveness, primarily through increasing directors' attention to cybersecurity issues and enabling them to ask incisive questions of management. Some findings:

- ...the overall consensus was that expertise enables directors to provide proactive, value-added oversight of cybersecurity risk that wouldn't be possible without it.
- [L]ack of expertise leads to superficial, check-the-box oversight. For example, board members may simply not give adequate attention to cybersecurity, since directors naturally focus on things they know best. They may ask the CISO naive or off-the-shelf questions that don't cut to the heart of the company's cybersecurity risks.
- When answers are provided, directors may not understand them, or be able to detect rosy framing or ask follow-up questions to probe if the CISO or their program needs a shift in direction.
- Benefits of directors with cybersecurity expertise include that they "can talk with the CISO to fully understand issues and challenge the CISO's cybersecurity programs when needed". They can also be a tremendous asset to the CISO by acting as a go-between with the board and by lending political capital to the CISO's requests to the C-suite for resources and needed organizational changes.
- [T]rue expertise — such as that gained through direct, hands-on experience managing cybersecurity — provides the most benefits. However, even modest investments in a board's expertise, such as cybersecurity training, can pay substantial dividends.

Footnotes

13. U.S. Senators Jack Reed, Mark Warner, Catherine Cortez Masto, Kevin Cramer, Susan Collins, Angus King, Ron Wyden. February 8, 2022. https://www.sec.gov/comments/s7-09-22/s70922-20127532-288660.pdf

COLUMBIA UNIVERSITY
US News and World Report #12 in National Universities

What They Said: We absolutely believe it is important to disclose the cyber expertise of the board to meet the SEC's three stated roles.

However, we believe, it is equally important to not require the expertise of particular individuals or their roles in cyber specific activities be disclosed. Making such disclosure voluntary is necessary to protect both the organization and promote access to board members of the highest quality. Disclosing specific expertise or roles of board members could make them and their families targets of nefarious activities such as extortion and doxing. Hence, the organization and/or the individuals may choose to make this level of disclosure, but it should not be mandated.[14]

TEXAS A&M UNIVERSITY SCHOOL OF LAW, and
US News and World Report: #26 Best Law Schools

PRAIRIE VIEW A&M UNIVERSITY
US News and World Report: #395 in National Universities

What They Said: It is a duty and responsibility of corporate directors to govern cybersecurity and cyber risk. Publicly traded corporations have a duty to disclose the existence of a data breach based upon at least two distinct authorities: Delaware common law and the SEC's 2011 corporate finance disclosure guidance, which identifies material data security risks that companies must disclose under securities law disclosure requirements and accounting standards. Accordingly, companies that know about a data breach but fail to disclose it to shareholders, regulators, and consumers, risk potential liability under corporate, breach notification, and securities laws.[15]

MICHIGAN STATE UNIVERSITY ELI BROAD COLLEGE OF BUSINESS, and
US News and World Report: #41 in Best Business Schools

UNIVERSITY OF ARIZONA DHALIWAL-REIDY SCHOOL OF ACCOUNTANCY
US News and World Report: #50 in Best Business Schools

What They Said: The findings of Ashraf et al. (2020) suggest that requiring an information technology or cybersecurity expert on the board may ultimately benefit shareholders.[16]

Who Is AGAINST Director Cyber Expertise Disclosure, And Why?

The AGAINST cohort reflects legacy governance associations, the American Bar Association (ABA) along with several law firms, Microsoft, a large number and swath of industry/trade associations and several others.

The arguments offered AGAINST director cyber expertise disclosure are composed of evidence free fallacies, anecdotes, and dismissive comments based upon illogical and sometimes unrelated issues or claims. Unlike the FOR cohort, not one organization offered any empirical evidence supporting their objections or the negative effects that would accrue as a result of disclosure (or presence on a board) of director cyber expertise.

It was frequently noted that disclosure would probably create a de facto standard that would "force" cybersecurity experts into the boardroom, a consequence that we believe would have also occurred and one that we support.

There is some language that was copied verbatim between the U.S. Chamber of Commerce comment letter and a blanket response on behalf of 34 different industry/trade associations. This demonstrates a coordinated propaganda campaign that was not independently or critically derived to advocate/lobby against this proposed provision.

It would be accurate to say that the submissions of the AGAINST cohort lack evidence, critical thinking, and rely heavily on unfounded and unproven declarations, or in other words, fear, uncertainty and doubt. Definitions for each of the identified FALLACIES are at the end of the paper.

Footnotes

14. Rajgopal, Shiva. Sharpe, Alex. Re: File Number S7-09-22, Comments on the SEC's Cybersecurity Risk Management, Strategy, Governance, and Incident Disclosure. Columbia University Graduate School of Business. May 9, 2022. https://www.sec.gov/comments/s7-09-22/s70922-20128410-291323.pdf

15. Trautman, Lawrence. Newman, Neal. A PROPOSED SEC CYBER DATA DISCLOSURE ADVISORY COMMISSION

16. Sunder, Dr. Jayanthi et al. Re: File number S7-09-22 – Cybersecurity Risk Management, Strategy, Governance, and Incident Disclosure. Michigan State University, The University of Arizona. May 8, 2022. https://www.sec.gov/comments/s7-09-22/s70922-20128266-290115.pdf

WHO IS AGAINST DIRECTOR CYBER EXPERTISE DISCLOSURE?
Corporate Governance Associations – National Association for Corporate Directors (NACD), Society for Corporate Governance
Industry Associations – US Chamber Of Commerce, Alliance For Automotive Innovation, American Petroleum Institute (API), Association Of Oil Pipe Lines (AOPL), American Fuel & Petrochemical Manufacturers (AFPM), American Chemistry Council, American Bar Association, Bank Policy Institute (BPI), American Bankers Institute (ABI), Independent Community Bankers Of America (ICBA), Mid-Size Bank Coalition Of America (MBCA), Biotechnology Innovation Organization (BIO), Business Roundtable, USTelecom – The Broadband Association, Canadian Bankers Association, Canadian Bankers Association, Consumer Technology Association, CTIA, Energy Infrastructure Council, Insurance Coalition, National Retail Foundation, National Restaurant Association, National Association Of Manufacturers, National Defense Industrial Association, R Street Institute, Real Estate Roundtable, ACA Connects—America's Communications Association, ACT | The App Association, Agricultural Retailers Association (ARA), Airlines for America (A4A), American Council of Engineering Companies (ACEC), American Council of Life Insurers (ACLI), American Gas Association (AGA), American Property Casualty Insurance Association (APCIA), Competitive Carriers Association (CCA), Consumer Technology Association (CTA), Federation of American Hospitals, The Fertilizer Institute (TFI), Global Business Alliance, Healthcare Information and Management Systems Society (HIMSS), Information Technology Industry Council (ITI), Interstate Natural Gas Association of America (INGAA), National Association of Broadcasters, National Association of Mutual Insurance Companies (NAMIC), National Association of Chemical Distributors (NACD), NCTA-The Internet & Television Association, National Electrical Manufacturers Association (NEMA), NTCA-The Rural Broadband Association, Professional Services Council (PSC), Semiconductor Industry Association (SIA), Telecommunications Industry Association (TIA)
IT/Cybersecurity Software Companies – Microsoft, Safe Securities Inc.
Law Firms – Davis Polk, Hunton Andrews Kurth, Wilson Sonsini
Accountants – BDO, Crowe, Deloitte
Others – NYSE, PPG

CORPORATE GOVERNANCE ASSOCIATIONS

NATIONAL ASSOCIATION OF CORPORATE DIRECTORS (NACD)

NACD is not-for-profit association for board directors. NACD is a collective of over 23,000 individuals, 1,750+ boards, more than 20 chapters across 35 locations.

What They Said: Disclosure of Director Expertise – As the proposed rule was written in 2022, we shared concerns that the new requirements would result in a one-size-fits-all approach, with all boards seeking to recruit single-issue directors, on whom they would (over) rely for cyber-risk oversight. As articulated in our NACD comment letter to the SEC, "The presence of a cybersecurity expert on a board cannot make up for poor oversight processes and does not excuse the full board from its oversight duties in this matter."

NACD believes it is regulatory overreach to impact the composition and operations of a public company board in this particular risk area, where effective governance is context dependent. NACD is pleased to see that the amended and final rule focuses disclosure of expertise at the correct level of responsibility for mitigating this risk: management.[17]

DDN Comments: The NACD statement that the presence of a cybersecurity expert on a board not being able to act as mitigating control for other poor oversight processes is wrong — it is the ultimate mitigating control. No examples are offered as to what they are referring to, but presumedly these other poor oversight processes could include audit committee responsibility and a lack of understanding of core cybersecurity risk issues such as systemic risk — but that is exactly what director cyber expertise does as the single most effective control to strengthening boardroom oversight.

Footnotes

17. Gleason, Peter. NACD RESPONDS TO SEC CYBER-RISK OVERSIGHT RULE. NACD Press Release. July 26, 2023.

Director cyber expertise is the one overriding mitigating control over any other oversight control weaknesses, e.g., audit committee alignment. Empirical evidence based upon actual experience supports the positive benefits in the research done in the FOR cohort by Virginia Tech and others.

In their SEC Comment letter, the NACD did make/include the following statement which contradicts their more current press release of July 2023 reflected above.

> We support identifying by name any directors with cybersecurity expertise (assuming the proposed safe harbor vii clause stays), and indeed would expand the disclosure to include statements on any directors who have received education specific to cyber-risk oversight best practices.

It is unclear why there is this conflicting statement on these issues from the NACD however their Press Release is their latest comment on this issue. It should be noted the NACD relies on third-parties to author some of their core published content on this issue, e.g., the NACD Director's Handbook on Cyber-Risk Oversight. This actually demonstrates the oft cited fallacy of relying heavily/exclusively upon "outside experts" in the boardroom, i.e., the OUTSIDE EXPERT FALLACY. This fallacy is used frequently in the AGAINST cohort to justify their position against director cybersecurity expertise.

This fallacy believes that boards can rely exclusively on outside experts, or even management, for governing cybersecurity. The OUTSIDE EXPERT FALLACY fails to acknowledge that director fiduciary duties cannot be delegated to a third-party and relies exclusively on outside experts rendering a periodic opinion or view at the behest of the board.

A board with low to no ability to understand and interpret the expert opinion, creates an exclusive, and often over reliance on the supposed authority. This is also a big reason why director financial expert disclsoure was a critical part of the Sarbanes-Oxley (SOX) regulations in 2022.

SOX in and of itself is evidence that the outside expert model has been proven to not work in corporate governance. The overreliance on "outside experts" leading up to the financial reporting crisis was a foundational cause behind the need for SOX and why the SEC regulated financial expertise into the boardroom in the first place. The boardroom needed the regular accountability that comes along with director financial expertise in the boardroom, as they do now with cybersecurity expertise. Research done by the FOR camp on this issue also demonstrates the weaker risk outcomes of a board incapable of understanding cybersecurity.

The ONE-SIZE FITS ALL FALLACY that the NACD argues in their press release is repeated by others in the AGAINST camp and is a popular misleading and dismissive argument related to this issue. The SEC proposed cybersecurity disclosure rule did not mandate a forced standard or definition of cybersecurity expertise. It did not require director cyber expertise on boards at all; it was merely a disclosure rule. Admittedly, that was a probable effect of the rule, but the SEC proposal did not explicitly define requirements for what constitutes cybersecurity expertise as this fallacy likes to represent.

FALLACIES PRESENTED				
ONE SIZE FITS ALL	ONE TRICK PONY	SCARCITY	SLIPPERY SLOPE	OUTSIDE EXPERT
✔				✔

SOCIETY FOR CORPORATE GOVERNANCE

As a not-for-profit organization (Section 501©(6)), the Society's role is to serve its members by providing practitioners with the knowledge, skills, and tools to promote effective governance for the benefit of boards, management, and shareholders. Founded in 1946, the Society for Corporate Governance, Inc. (the "Society") is comprised principally of corporate secretaries and business executives in governance, ethics and compliance functions at public, private and not-for-profit organizations.

What They Said: We believe this specific expertise disclosure mandate will decrease, rather than increase, the effectiveness of board oversight of cybersecurity, as well as its oversight of other issues. While it is important for boards to have a mix of directors with different skills and qualifications, those skills are generally considered broadly – companies often disclose that their directors have experience with areas like "marketing" or "international business," for instance, rather than trying to closely tie such skills to the specific roles, work experiences, and educational background. We believe this is appropriate for the role that directors play – a board of directors should not operate as a second management team, with individual directors replicating the most important management roles. Rather, a board exists to provide high-level oversight of strategy and operations, helping ensure management is balancing priorities, taking advantage of opportunities, and protecting against risks.

In 2018, when the U.S. Senate was considering the Cybersecurity Disclosure Act, which would have legislatively required issuers to disclose whether their boards contained cybersecurity experts, our CEO, Darla Stuckey, testified to the Banking Committee about the risks of appointing "special interest" directors to boards. During her testimony, she quoted a Society member:

> Of course, we want our public companies' boards to have the requisite skills to deal with all sorts of issues. However, specifying the types of skills that a company's board must have strikes me as the ultimate one-size-fits-all approach and has no logical limits. Should every public company have an expert on revenue recognition? Related party transactions? Has anyone thought through the consequences of having a board comprised of one-issue experts who may not have any other applicable skill sets? And would a cyber-expert want to be on a board, given that he or she would likely be blamed (and possibly sued) if the company had a breach or other cyber problem?[18]

DDN Comments: The ONE TRICK PONY FALLACY dismissively pigeon-holes cybersecurity experts as too narrow and incapable of contributing to any other boardroom agenda item and is a common and oft repeated fallacy. One which we believe demonstrates a lack of understanding of the CISO role, fails to acknowledge directors with financial expertise and their functional background and expertise along with any other functional depth naturally resident in any other corporate director, e.g., marketing.

The SCG comments about the board potentially being a second management team if there are directors with deep expertise on issues is also a false premise and a SLIPPERY SLOPE FALLACY. For starters, under Delaware law companies can be "managed by or under the direction of the board." This boundary is necessarily broad and leaves the details of how to do govern largely up to the board itself. The level of engagement and role of the corporate board is dictated by needs, circumstance and what the board/management team determine and make it to be. To say that a board of directors that knows too much is somehow a bad thing for corporate oversight fails to recognize that the board is an integral part of the system, not removed from it.

They also included their CEO's congressional commentary which includes the common ONE SIZE FITS ALL FALLACY that the proposed SEC rule would force the same standard into place for all issuers, which it does not. Neither the NACD nor The Society for Corporate Governance offered any empirical evidence to support their objections.

FALLACIES PRESENTED				
ONE SIZE FITS ALL	ONE TRICK PONY	SCARCITY	SLIPPERY SLOPE	OUTSIDE EXPERT
✔	✔		✔	

INDUSTRY ASSOCIATIONS

U.S. Chamber of Commerce

The Chamber of Commerce of the United States is the world's largest business organization. Their members range from the small businesses and chambers of commerce across the country that support their communities, to the leading industry associations and global corporations that innovate and solve for the world's challenges, to the emerging and fast-growing industries that are shaping the future.

What They Said: Company boards should prioritize managing cyber risks but not through SEC mandates requiring cybersecurity "expertise". Our associations advocate for companies to proactively prioritize cyber risk management activities, but they are concerned about the SEC's call for companies to disclose the

Footnotes

18. Allen, Ted. Stuckey, Darla. RE: Comment Letter on Proposed Rule: Cybersecurity Risk Management, Strategy, Governance, and Incident Disclosure, SEC File Number S7-09-22.
Society for Corporate Governance. May 9, 2022. https://www.sec.gov/comments/s7-09-22/s70922-20135637-306403.pdf

name of any board member who has cybersecurity expertise. We believe that board experts should not proliferate via government directives. Prescriptive disclosures intended to drive company behavior regarding which subject-matter experts sit on companies' governing bodies could lead to unwieldy and unwanted outcomes (e.g., giving investors a false sense of confidence because of the presence of a board cybersecurity "expert").

Also, cybersecurity talent is scarce globally. It is unclear where companies would get the cybersecurity experts that would be driven by the Commission's proposed requirement to disclosure such expertise. There is a well-established lack of cybersecurity talent for the public and private sectors that would impede companies' abilities to recruit board cybersecurity experts. The SEC's proposal could even create unintended barriers for historically under-represented groups to move into cybersecurity management or board leadership roles — not due to the lack of qualifications but to the absence of formal credentials (e.g., owing to their costs) and other certifications. Even if companies could obtain the relevant cybersecurity experts for board positions, no evidence has been convincingly shown that this requirement would inform investors or improve companies' cyber-security preparedness.[19]

DDN Comments: The U.S. Chamber of Commerce offers no evidence to support their claims. They have presented the claim of "unwieldy and unwanted outcomes" if disclosure would be required which is the SLIPPERY SLOPE FALLACY of bad things to come and the false sense of security that could be created if in fact there were director cyber expertise on the board, which fails basic logic. This is also illogical as it is similar to arguing that having a security guard is worse than having no guard at all in terms of the overall security profile within a bank.

They also deploy the SCARCITY FALLACY to argue that there are not enough potential director cyber experts to go around, so why even bother. Evidence was presented by the FOR cohort to the contrary.

FALLACIES PRESENTED				
ONE SIZE FITS ALL	ONE TRICK PONY	SCARCITY	SLIPPERY SLOPE	OUTSIDE EXPERT
		✔	✔	

ALLIANCE FOR AUTOMOTIVE INNOVATION

Auto Innovators is the singular, authoritative, and respected voice of the automotive industry. Focused on creating a safe and transformative path for personal mobility, Auto Innovators represents the manufacturers that produce nearly 98 percent of cars and light trucks sold in the United States, original equipment suppliers, technology companies, and other value chain partners within the automotive ecosystem.

What They Said: Disclosing whether a registrant's Board of Directors has specific cybersecurity expertise also has little bearing on the overall cybersecurity posture of a firm, especially when a firm may have other personnel, such as a Chief Information Security Officer or CISO, who directs day-today cybersecurity risk management operations.[20]

DDN Comments: A twist on the OUTSIDE EXPERT FALLACY, they are arguing that cybersecurity governance would not strengthen the organization's system of cybersecurity saying it is up to management. Failing to understand the board is a control in the cybersecurity system, and not understanding the purpose of corporate governance was not a condition for submitting SEC comments.

FALLACIES PRESENTED				
ONE SIZE FITS ALL	ONE TRICK PONY	SCARCITY	SLIPPERY SLOPE	OUTSIDE EXPERT
				✔

AMERICAN PETROLEUM INSTITUTE (API), ASSOCIATION OF OIL PIPELINES (AOPL), AMERICAN FUEL & PETROCHEMICAL MANUFACTURERS (AFPM)

These organizations represent companies from all segments of the American oil and natural gas industry and are involved in exploration, production, refining, marketing, distribution, and marine activities.

What They Said: The Proposal would effectively impose prescriptive one-sized-fits-all mandates with respect to board-member expertise with

Footnotes

19. Quaadman, Tom et al. Re: Cybersecurity Risk Management, Strategy, Governance, and Incident Disclosure (File Number S7-09-22). U.S. Chamber of Commerce. May 9, 2022. https://www.sec.gov/comments/s7-09-22/s70922-20149099-316288.pdf

20. Hairston, Tara. RE: Cybersecurity Risk Management, Strategy, Governance, and Incidence Disclosure Proposed Rule (File Number S7-09-22). Auto Innovators. May 9, 2022. https://www.sec.gov/comments/s7-09-22/s70922-20128350-291113.pdf

cybersecurity issues. To be sure, board members must be aware of cybersecurity threats to the company like they do for other important risks. For example, directors should understand company cybersecurity vulnerabilities and should be able to drive decisions regarding a company's overall efforts to guard against and address those risks. However, it is not clear why specific director expertise on cyber issues is necessary or relevant for all issuers and there should not be an expectation that most companies maintain a board of directors that has such an expert in cybersecurity — or that a company should have to explain why it doesn't have an expert on its board.

Board members do not need expertise in reverse engineering malware, the tactics, techniques, and procedures of foreign intelligence services and cybercriminals, or other highly technical computer network operation topics. Instead, most registrants, and especially all large ones, will have a Chief Information Security Officer and significant access to technical expertise that can inform the board on addressing these issues and situations. In addition, the Proposal does not adequately describe how to even measure cybersecurity expertise among the Board or why having board members with technical expertise regarding cyber risks would be more beneficial than having other cybersecurity resources within a company. For example, individual technical degrees may not include the practical knowledge of leading teams that address current cybersecurity threats and incidents. Accordingly, the benefit case for needing cybersecurity-specific expertise on the Board, or requiring an explanation of why not, does not appear to be clearly articulated by the Proposal.[21]

FALLACIES PRESENTED				
ONE SIZE FITS ALL	ONE TRICK PONY	SCARCITY	SLIPPERY SLOPE	OUTSIDE EXPERT
✔				✔

AMERICAN CHEMISTRY COUNCIL

The American Chemistry Council (ACC) represents more than 190 companies engaged in the business of chemistry — an innovative, economic growth engine that is helping to solve the biggest challenges facing our country and the world.

What They Said: Cybersecurity talent is hard to find. From a personnel standpoint, it's unclear where companies would get the so-called cybersecurity expertise that the proposed regulation would mandate. There is a well-known lack of cybersecurity talent for the public and private sectors. In any case, the SEC has not justified that having cybersecurity experts on boards would benefit its cybersecurity protection program. Similarly, investors may see the inclusion of certain individuals on a company's board as an indication of a company's overall cybersecurity program maturity and as a sign that a company is more secure than another one. Such an outcome could be misleading.[22]

DDN Comments: Making an appearance in these arguments is the SCARCITY FALLACY where there aren't enough cyber experts and also the SLIPPERY SLOPE FALLACY where a director cyber expert may make a company more secure, relative to others, but it could be misleading, although they do not explain how and the argument falls apart logically.

FALLACIES PRESENTED				
ONE SIZE FITS ALL	ONE TRICK PONY	SCARCITY	SLIPPERY SLOPE	OUTSIDE EXPERT
		✔	✔	

AMERICAN BAR ASSOCIATION

The ABA was founded in 1878 on a commitment to set the legal and ethical foundation for the American nation. Today, it exists as a membership organization and stands committed to its mission of defending liberty and pursuing justice.

DDN Comment: The ABA's response on ABA letterhead was accompanied by this caveat — The comments expressed in this letter represent the views of the Committee only and have not been approved by the ABA's House of Delegates or Board of Governors and, therefore, do not represent the official position of the ABA. In addition, this letter does not represent the official position of the Business Law Section of the ABA, nor does it necessarily reflect the views of all members of the Committee.

Footnotes

21. Comstock, Stephen, et al. File Number S7-09-22 API, AOPL, AFPM. May 9, 2022. https://www.sec.gov/comments/s7-09-22/s70922-20128380-291283.pdf

22. Gulledge, Bill. Re: Cybersecurity Risk Management, Strategy, Governance, and Incident Disclosure (File Number S7-09-22). ACC. November 1, 2022. https://www.sec.gov/comments/s7-09-22/s70922-20149102-316290.pdf

What They Said: The Committee believes that a board's role in risk oversight, including cybersecurity risk oversight, does not require members of the board to have specific expertise in cybersecurity. Companies retain and benefit from employees or advisors with the cybersecurity expertise to manage and advise on highly technical cybersecurity topics or areas of concern applicable to the company and its industry. The board's role is to oversee an ever-changing range of risks confronting a company, which requires broad-based skills in risk and management oversight, rather than subject matter expertise in one particular type of risk.

Consequently, requiring specific disclosure regarding a director's expertise in cybersecurity may have the unintended effect of boards de facto delegating cybersecurity risk oversight to the identified individual(s) and reducing the board's sense of collective responsibility for cybersecurity risk oversight. There is also great potential for investors to draw the wrong conclusion about the effectiveness of a board's oversight of cybersecurity based on the identification (or not) of a director with cybersecurity expertise.[23]

DDN Comments: ABA presents the arguments that management and/or outside experts are sufficient for cybersecurity governance, i.e., the OUTSIDE EXPERT FALLACY. And they also go to the SLIPPERY SLOPE FALLACY with the argument that a full board with no cybersecurity expertise on it, but collective responsibility is a better outcome than having a competent point of accountability on the board for cybersecurity oversight. This is like a plane without a pilot relying upon the hope that the passenger's collective responsibility can figure out how to fly and land the plane.

This is accompanied by a comment about investors drawing the wrong conclusion about affirmative disclosure or negative disclosure of director cybersecurity expertise, offering an argument that no disclosure is equivalent to disclosure and that investors tare not able to distinguish which is more effective — investors are in the FOR camp.

FALLACIES PRESENTED				
ONE SIZE FITS ALL	ONE TRICK PONY	SCARCITY	SLIPPERY SLOPE	OUTSIDE EXPERT
			✔	✔

BANK POLICY INSTITUTE (BPI), AMERICAN BANKERS INSTITUTE (ABA), INDEPENDENT COMMUNITY BANKERS OF AMERICA (ICBA), MID-SIZE BANK COALITION OF AMERICA MBCA)
BPI is a nonpartisan group representing the nation's leading banks. Collectively, BPI members hold $10.7 trillion in deposits in the United States, make 68% of all loans, including trillions of dollars in funding for small businesses and household mortgages, credit cards, and auto loans; employ nearly two million Americans, and serve as a principal engine for the nation's financial innovation and economic growth. ABA is the voice of the nation's $23.3 trillion banking industry, which is composed of small, regional and large banks that together employ millions of people, safeguard $19.2 trillion in deposits and extend nearly $11 trillion in loans. ICBA creates and promotes an environment where community banks flourish. MBCA's member banks average less than $20 billion in size and serve customers and communities through more than 10,000 branches in all 50 states, the District of Columbia, and three U.S. territories.

What They Said: Finally, the Associations do not support a new requirement that registrants disclose the cybersecurity expertise of members of a board of directors. The proposed requirement will have the effect of suggesting that boards without directors with such specific expertise are somehow deficient. Pressure from the SEC for public companies to designate directors with expertise in any single area may adversely impact their ability to identify and appoint directors with other experience, expertise, or capabilities they believe are appropriate for the oversight of the particular risks and opportunities the institution encounters and may not result in more effective cyber risk oversight for the institution.

While the Associations agree that cybersecurity is a significant risk that requires appropriate board oversight, we believe the proposed disclosure requirement is problematic. The proposed requirement will have the effect of suggesting that boards without directors with such specific expertise are somehow deficient, and it comes in the context of other proposed requirements that have the

Footnotes

23. Knight, Jay. File No. S7-09-22 Release Nos. 33-11038; 34-94382 Cybersecurity Risk Management, Strategy, Governance, and Incident Disclosure. ABA. July 20, 2022. https://www.sec.gov/comments/s7-09-22/s70922-20134430-304137.pdf

overall effect of encouraging companies to supplant directors with expertise overseeing diverse risks and complex institutions with those who have single subject matter expertise akin to that of senior managers.[24]

DDN Comments: The Associations make the argument that the lack of a control, i.e., boardroom cybersecurity expertise, infers a deficiency. Which is exactly right. It doesn't just infer a control gap, it confirms it. They also present an evidence free argument on the ONE TRICK PONY FALLACY that having cybersecurity expertise somehow creates a mutual exclusivity with the inclusion or needs for other functional or domain expertise in the boardroom.

FALLACIES PRESENTED				
ONE SIZE FITS ALL	ONE TRICK PONY	SCARCITY	SLIPPERY SLOPE	OUTSIDE EXPERT
	✔		✔	

BIOTECHNOLOGY INNOVATION ORGANIZATION (BIO)

BIO is the world's largest life sciences trade association representing nearly 1,000 biotechnology companies, academic institutions, state biotechnology centers and related organizations across the United States and in more than 30 other nations.

What They Said: The Commission is adding to this complexity [the general difficulty in finding qualified directors already in the sector and for small business] with the need for cybersecurity expertise.

Put another way, **cyberattacks mainly affect large companies and are not material for smaller companies**. (bolding is theirs). As we note above and is implied in the Commission's proposed rule and citations within the proposed rule, smaller companies are not the main targets for cybercriminals. Hence, the proposed Item 407(j) seems more appropriate as a requirement for index inclusion.

[Specifically with regard to named disclosure] The cybercriminal industry prides itself on public embarrassment. There is little benefit in providing fodder to those with malicious intent.

Singling out a person is never a good idea and will bring unwanted attention to a company's oversight structure. How companies handle cybersecurity at the board level need not be detailed.[25]

DDN Comments: Declaring that small companies are not targets for cybersecurity demonstrates a lack of understanding and is incorrect. Almost warranting its own fallacy is the name/shame statement of singling out an expert. Something that already occurs with financial expertise or any other statement or identification of director skills which is already made for public company directors in proxy disclosures.

FALLACIES PRESENTED				
ONE SIZE FITS ALL	ONE TRICK PONY	SCARCITY	SLIPPERY SLOPE	OUTSIDE EXPERT
		✔	✔	

BUSINESS ROUNDTABLE

Business Roundtable is an organization whose members lead America's largest companies, employing over 20 million workers. Their companies' total value, over $20 trillion, accounts for half of the value of all publicly traded companies in the United States.

What They Said: We agree that disclosure about the skills, background and expertise of board members is important information for our investors, along with disclosure of oversight regarding risk management. In fact, many of our member companies already provide disclosure regarding the cybersecurity expertise of board members and how boards oversee, and company leadership manages, cybersecurity risk. That said, we caution against an ever-expanding set of disclosure requirements regarding specific skills without regard to the materiality of that particular skill to the registrant. Further, we do not believe that the proposed disclosure requirements regarding cybersecurity expertise are necessary at this time. The proposed rules would require a registrant not only to disclose whether any directors have expertise in cybersecurity, but also to "provide such detail as necessary to fully describe the nature of the expertise."

Footnotes

24. Feeney, Christoper et al. Proposed Rules Regarding Cybersecurity Risk Management, Strategy, Governance, and Incident Disclosure Requirements (File No. S7-09-22). BPI, ABA, ICBA, MBCA, May 9, 2022. https://www.sec.gov/comments/s7-09-22/s70922-20128336-291093.pdf

25. Passeri, Carlo. Re: File No. S7-09-22: Cybersecurity Risk Management, Strategy, Governance, and Incident Disclosure. BIO. May 5, 2022. https://www.sec.gov/comments/s7-09-22/s70922-20128401-291312.pdf

Companies do not need directors who are experts at forensic data analysis or similar technical aspects of incident response. Instead companies benefit from directors who have significant leadership and risk management experience and have demonstrated a capacity to learn new skills and expertise. Otherwise, the SEC risks driving companies to create boards filled with "specialty directors" who have deep but narrow knowledge and struggle to fulfill the broad oversight and related duties required today.[26]

FALLACIES PRESENTED				
ONE SIZE FITS ALL	ONE TRICK PONY	SCARCITY	SLIPPERY SLOPE	OUTSIDE EXPERT
	✔			

USTELECOM – THE BROADBAND ASSOCIATION

USTelecom members are technology innovators creating new opportunities and connections for families, enterprises, and communities.

What They Said: USTelecom would support general reporting about risk management backgrounds, but narrowing in on cybersecurity expertise does not take into account the diverse backgrounds and experiences that will contribute to a successful and diverse board.[27]

FALLACIES PRESENTED				
ONE SIZE FITS ALL	ONE TRICK PONY	SCARCITY	SLIPPERY SLOPE	OUTSIDE EXPERT
	✔			

CANADIAN BANKERS ASSOCIATION

The CBA is a professional industry association that provides information, advocacy education and operational support services to its membership of more than 60 domestic and foreign banks operating in Canada.

What They Said: We believe that this disclosure requirement would exert pressure on registrants to appoint directors with such expertise, which may come at a significant cost to the ability of their boards to oversee other risks to the company without proportionate benefits, particularly at boards of heavily regulated financial institutions. Further, by appointing directors with single subject-matter expertise, registrants may experience less effective board oversight, as those with "expertise" may naturally assume responsibility over cybersecurity risk and management, which the board as a whole should oversee. The proposed responsibility allocated to the board could blur the lines between the typical or existing responsibility allocation between the board and management.[28]

FALLACIES PRESENTED				
ONE SIZE FITS ALL	ONE TRICK PONY	SCARCITY	SLIPPERY SLOPE	OUTSIDE EXPERT
	✔		✔	

CENTER FOR AUDIT QUALITY

The Center for Audit Quality (CAQ) is a nonpartisan public policy organization serving as the voice of U.S. public company auditors and matters related to the audits of public companies.

What They Said: Expertise is a higher bar than having sufficient knowledge and understanding of a subject matter. In many cases, boards of directors having sufficient knowledge and understanding of cybersecurity risks (and any other evant risks) facing the registrants they oversee, rather than specific expertise in one area of focus, gives them the ability to execute appropriate oversight activities. Depending on the cybersecurity risk profile of an organization, a board may be able to obtain relevant cybersecurity knowledge and affirm understanding by having expert individuals within an organization through periodic training, knowledge-sharing, and other activities.

While board-level cybersecurity expertise may be appropriate for certain registrants, for others the board can execute reasonable cybersecurity risk management oversight without having such a specific expertise. It should be at the discretion of registrants to determine the level of board cybersecurity expertise required and the manner which boards obtain such expertise. As boards of directors are generally limited in capacity, requiring disclosure of board cyber expertise may result in boards shifting their composition to prioritize cyber

Footnotes

26. Feeney, Christoper et al. Proposed Rules Regarding Cybersecurity Risk Management, Strategy, Governance, and Incident Disclosure Requirements (File No. S7-09-22). BPI, ABA, ICBA, MBCA, May 9, 2022. https://www.sec.gov/comments/s7-09-22/s70922-20128336-291093.pdf

27. Passeri, Carlo. Re: File No. S7-09-22: Cybersecurity Risk Management, Strategy, Governance, and Incident Disclosure. BIO. May 5, 2022. https://www.sec.gov/comments/s7-09-22/s70922-20128401-291312.pdf

28. Hannah, Darren. Cybersecurity Risk Management, Strategy, Governance, and Incident Disclosure (File No. S7-09-22). CBA. May 5, 2022. https://www.sec.gov/comments/s7-09-22/s70922-20128288-290991.pdf

expertise over other important areas of focus to the detriment of overall corporate governance.[29]

DDN Comments: Arguing that director cyber expertise creates a mutual exclusivity with other boardroom needs creates a false SLIPPERY SLOPE.

FALLACIES PRESENTED				
ONE SIZE FITS ALL	ONE TRICK PONY	SCARCITY	SLIPPERY SLOPE	OUTSIDE EXPERT
		✔	✔	✔

CONSUMER TECHNOLOGY ASSOCIATION

The Consumer Technology Association (CTA)® convenes companies of every size and specialty in the technology industry to move us all forward. CTA is the trade association representing the $505 billion U.S. consumer technology industry, which supports more than 18 million U.S. jobs.

What They Said: While CTA agrees that cybersecurity already is and should always be among the top priorities for most public companies, "cybersecurity expertise" by its nature is diverse — it can be both sector-specific and dependent on the innovative capabilities of interdisciplinary experts — and likely cannot be adequately quantified in an SEC disclosure. Such a requirement would more likely push companies to "check the box" with token qualifications among one or more board members, rather than a discerning, holistic suite of risk management experience and expertise — including diverse cybersecurity experience and expertise — that companies apply in their board appointments.[30]

DDN Comments: Their statement recognizes the value and diverse breadth of cybersecurity expertise on one hand, and then heads down an illogical SLIPPERY SLOPE of this becoming a mere compliance exercise.

FALLACIES PRESENTED				
ONE SIZE FITS ALL	ONE TRICK PONY	SCARCITY	SLIPPERY SLOPE	OUTSIDE EXPERT
			✔	

CTIA

CTIA represents the U.S. wireless communications industry and the companies throughout the mobile ecosystem that enable Americans to lead a 21st-century connected life.

What They Said: The SEC proposes to require public companies to disclose the cyber expertise of the members of the board of directors, with the apparent goal of encouraging companies to select board members with cybersecurity backgrounds or experience. This requirement should not be adopted. The criteria outlined by the SEC in determining whether a board member has "cybersecurity expertise" does not reflect the range of experiences, beyond cybersecurity backgrounds or credentials, that could contribute to robust board oversight of cyber risk management. Board members may have varied experiences that contribute to oversight of a company's cyber risk, such as having managed organizations through cyber response in the past.

Companies pursue various goals in building diverse and innovative boards; these efforts, which benefit investors, should not be limited or disfavored by additional expectations for specialized expertise that may not be plentiful across the economy.

The Commission's proposal could crowd out other areas of expertise and harm companies that provide critical services to the U.S. government. Smaller or resource constrained companies may face challenges in hiring board members that have the specific cyber expertise the Commission would require and be put at a disadvantage in the market.[31]

DDN Comments: A few of the standard fallacies in here and their last statement implies they believe broad based cybersecurity weakness is preferable if smaller firms cannot bring on a board cyber expert with the argument that if it can't be fixed for all, it should not be fixed for any.

FALLACIES PRESENTED				
ONE SIZE FITS ALL	ONE TRICK PONY	SCARCITY	SLIPPERY SLOPE	OUTSIDE EXPERT
	✔	✔	✔	

Footnotes

29. McGowan, Dennis. Re: File Number S7-09-22: Cybersecurity Risk Management, Strategy, Governance, and Incident Disclosure; Release Nos. 33-11038; 34-94382; IC-34529. Center for Audit Quality. May 9, 2022. https://www.sec.gov/comments/s7-09-22/s70922-20128390-291293.pdf

30. Grossman, J. David. Bergman, Mike. Proposed Rules for Cybersecurity Risk Management, Strategy, Governance and Incident Disclosure, File No. S7-09-22. CTA. May 9, 2022. https://www.sec.gov/comments/s7-09-22/s70922-20128344-291103.pdf

31. Tiano, Melanie et al. In the Matter of Cybersecurity Risk Management, Strategy, Governance, and Incident Disclosure. CTIA. May 9, 2022. https://www.sec.gov/comments/s7-09-22/s70922-20128384-291287.pdf

ENERGY INFRASTRUCTURE COUNCIL

The Energy Infrastructure Council ("EIC") is a non-profit trade association of companies that develop and operate energy infrastructure, including traditional and renewable energy infrastructure companies; investors in energy infrastructure; service providers; and other businesses and individuals that operate in and around the energy industry.

What They Said: Additionally, requiring the disclosure of cybersecurity expertise for a member of the board of directors may be difficult while providing limited benefit to investors. Boards of directors are distinct from management, as the board's role is one of oversight whereas management is required to have subject matter expertise. The board should have the flexibility to determine its own composition, and needs, and take into consideration the collective expertise of the board, holistically. Boards are, by design, deliberative bodies and tasked with oversight of numerous risks — of which cyber is only one of those risks. Current disclosures required concerning board's business experience should be sufficient to elicit relevant information for investors.[32]

INSURANCE COALITION

Insurance Coalition, a group of life and property and casualty insurance companies that share a common interest in federal regulations.

What They Said: In addition to unclear definitions of "expertise" and the associated compliance costs discussed above, disclosure of cybersecurity expertise without a mandate for cyber experts to serve on a Board, will necessarily signal potential vulnerabilities to cyber criminals — thereby exposing firms to even greater cyber risk.

Having cyber expertise on a public firm Board of Directors is a reasonable means of ensuring good corporate governance. Pursuing this through public, and not private, policy however necessarily creates a Catch-22. Requiring disclosure of cyber expertise on a Board of Directors without a mandate unnecessarily exposes vulnerabilities of firms without cyber expertise, or with limited expertise. This in turn makes reporting a de facto mandate for cyber experts to serve on a board for reasonable firms. Mandating cyber experts to serve on public company boards, however, is a potential overreach of the Commission as it denies firms the ability to compete on cybersecurity as a means of protecting shareholders, customers, and investors and ultimately, innovation in both cybersecurity hygiene and corporate governance.[33]

DDN Comments: This is another confusing one. On one hand there's an acknowledgment that not having a cyber expert on a board signals a weak cybersecurity environment to hackers and that having one would strengthen it. They also state that disclosure would be a de facto mandate, which they believe is overreach as it would deny firms the ability to "compete" on cybersecurity, which I must admit we don't understand.

FALLACIES PRESENTED				
ONE SIZE FITS ALL	ONE TRICK PONY	SCARCITY	SLIPPERY SLOPE	OUTSIDE EXPERT
			✔	

NATIONAL RETAIL FOUNDATION

NRF is the world's largest retail trade association, representing discount and department stores, home goods and specialty stores, Main Street merchants, grocers, wholesalers, chain restaurants, and internet retailers from the United States and more than 45 other countries.

What They Said: The governance disclosure rule is unnecessary to protect investors and micro-manages company cybersecurity governance measures, which goes beyond the Commission's expertise.

The SEC has not explained why a detailed list of governance-related cybersecurity disclosures is needed for the protection of investors. The regulation would require, among other things, the disclosure of: (1) a board member's cybersecurity expertise, (2) whether the company has a chief information security officer, their relevant expertise, and where they fit in the organizational chart, and (3) granular information about the interactions of management and the board

Footnotes

32. Ziebart, Lori. RE: File No. S7-09-22; RIN 3235-AM89: SEC Proposed Rule: Cybersecurity Risk Management, Strategy, Governance, and Incident Disclosure. EIC. May 9, 2022. https://www.sec.gov/comments/s7-09-22/s70922-20128379-291282.pdf

33. Ostro, Zach. File No.: S7-09-22. The Insurance Coalition. May 9, 2022. https://www.sec.gov/comments/s7-09-22/s70922-20128367-291142.pdf

of directors on cybersecurity, including the frequency with which the various actors discuss the topic. This level of detail in governance-related information is unprecedented in SEC regulation, is inflexible in its approach, and is simply not needed for investors to make sound investment decisions.

For instance, the proposal's requirement that companies disclose the cybersecurity expertise of board members (which implicitly preferences such expertise in board members) may be unsound business judgment for most registrants. Board members are not directly implementing cybersecurity policy — especially in industries such as retail, in which cyber security is not the core business — and it is unclear why there should be an expectation that all public companies search for and elect board members with cybersecurity expertise. Of course, board members should be aware of technology and general cybersecurity risks to help make decisions and manage corporate risk on behalf of the company. But board members are not directly implementing cybersecurity policy or responding to cybersecurity incidents, which makes specific training and expertise in cybersecurity response measures unnecessary. In addition, as a practical matter, because cybersecurity best practices and technical capabilities are constantly evolving, it is unclear how, for example, a decades-old degree or certification in cybersecurity would be especially helpful in managing the risks cybersecurity poses to public companies today.[34]

FALLACIES PRESENTED				
ONE SIZE FITS ALL	ONE TRICK PONY	SCARCITY	SLIPPERY SLOPE	OUTSIDE EXPERT
			✔	

NATIONAL RESTAURANT ASSOCIATION

Founded in 1919, the National Restaurant Association ("the Association") is the leading business association for the restaurant and food-service industry, representing more than 14.5 million employees, nearly 10 percent of the nation's workforce.

What They Said: The Association recognizes the importance of films maintaining robust governance structures and comprehensive compliance programs with a reporting line to escalate cyber issues to senior management and the Board (or Board Committee). We agree with the Commission that Boards should exercise some oversight of cybersecurity programs. However, governance structures are different for companies where cybersecurity is mission-critical (i.e., critical infrastructure entities) compared to companies that merely require good cyber hygiene to operate successfully on a day-today basis. Ultimately, the Association believes that the precision of the Board composition reporting requirement to disclose the specific degree of their involvement in approving policies and procedures is hard to quantify, overly prescriptive and difficult to implement in a practical way. Instead, we suggest that it may make sense for the Commission to tailor the Board requirements to CISA-aligned critical industries.[35]

NATIONAL ASSOCIATION OF MANUFACTURERS

The NAM is the largest manufacturing trade association in the United States, representing manufacturers of all sizes and in all 50 states.

What They Said: As a matter of first course, the NAM does not believe that it is appropriate for the SEC to mandate specific criteria for candidates to public company boards of directors. Company boards are charged with broad strategy, governance, and risk management duties, and issuers and their shareholders are free to nominate and elect directors with a mix of diverse experience and expertise in order to meet these challenges. The SEC should not require the consideration or election of certain types of director candidates or impose disclosure requirements designed to achieve similar goals. Specific to cybersecurity, manufacturers know that effective cybersecurity governance does not require a single director with a certain technical degree, but rather an enterprise-wide focus on the issue and appropriate risk prioritization by the board. Effective risk management, including with respect to cybersecurity, falls under the purview of the entire board and management team, not just one director.

Further, the proposed rule's criteria for cybersecurity expertise are unnecessarily limiting and could

Footnotes

34. Martz, Stephanie. RE: File Number S7-09-22, Comments of the National Retail Federation in Response to the SEC's Proposed Rule on Cybersecurity Risk Management, Strategy, Governance, and Incident Disclosure. NRF. May 9, 2022. https://www.sec.gov/comments/s7-09-22/s70922-20128322-291085.pdf

35. Duckett, Brennan. Re: National Restaurant Association Comments on 17 CFR Parts 229, 232, 239, 240, and 249 [Release Nos. 33-11038; 34-94382; IC-34529; File No. S7-09-22]. Association. May 9, 2022. https://www.sec.gov/comments/s7-09-22/s7092220128325-291052.pdf

discourage companies from considering otherwise qualified director candidates because their experience does not fit neatly under one of the SEC's proposed credentials. The qualifications necessary to manage cybersecurity risk in today's evolving threat landscape may not always align with the limited criteria in the proposed rule, and the rule's proposed qualifications could become more out-of-date as the threat landscape continues to evolve in the years to come. Further, the proposed rule focuses on specific technical expertise, job responsibilities, and professional certifications—yet the board of directors is charged with oversight and management of cybersecurity risk, not hands-on technology implementation and incident response. Expertise in these critical areas should also be prized by public companies looking for board leadership, not just specific technical skills.[36]

DDN Comments: Fails to recognize that the SEC has mandated director skills in the past, e.g., with financial expertise along with other significant corporate governance reforms with SOX. When an issue is so pervasive and embodies significant risks to the interests of investors and consumers, as cybersecurity does, the capability and expertise to universally strengthen the boardroom in its control function becomes universally warranted. We also believe the unique systemic nature of cybersecurity risk also justifies a collective requirement and standard, something unlikely to be achieved without regulatory action, i.e., your weakness and hack could be problems for many other companies like CDK is experiencing or CRWD.

FALLACIES PRESENTED				
ONE SIZE FITS ALL	ONE TRICK PONY	SCARCITY	SLIPPERY SLOPE	OUTSIDE EXPERT
	✔		✔	

NATIONAL DEFENSE INDUSTRIAL ASSOCIATION

The National Defense Industrial Association (NDIA) represents more than 1,800 corporate and over 60,000 individual members from small, medium, and large contractors, educational institutions, and governments.

What They Said: NDIA is concerned that it may not be essential that Board members themselves have specialized expertise in cybersecurity. There already is a shortage of personnel with cybersecurity expertise to address ongoing challenges and risks associated with cybersecurity and cybersecurity incident identification, reporting, and remediation. Public companies need appropriate oversight by Board members and can hire cyber expertise for specific roles within the company and / or as consultants to the Board where and when needed.[37]

FALLACIES PRESENTED				
ONE SIZE FITS ALL	ONE TRICK PONY	SCARCITY	SLIPPERY SLOPE	OUTSIDE EXPERT
	✔	✔		✔

NYSE

The NYSE is the world's largest exchange, home to more than 2,400 listed companies having an aggregate market capitalization of more than $35 trillion. In this capacity, the Exchange recognizes the critical role that technology plays for companies across all industries and sectors. Given the importance of technology to the business and operations of virtually all listed companies, the Exchange acknowledges that information around an issuer's cybersecurity risk management and incident response may be material to investors.

What They Said: Although these proposed amendments do not mandate that corporate boards include a member with expertise in cybersecurity, they have the implicit suggestion that corporate boards should include a member with such qualifications. Failure to have a board member with expertise in cybersecurity, therefore, could result in investors reaching the mistaken conclusion that a company is unconcerned with cybersecurity.

While the Exchange agrees with the Commission that cybersecurity is an important area of focus for nearly all public companies, it does not believe that the absence of a cybersecurity expert on a company's board is necessarily the fatal flaw that the required disclosure may implicitly suggest to investors. Corporate boards consider a wide range of important issues and should have broad discretion to determine how they educate themselves on each subject.

Footnotes

36. Netram, Chris. Re: File No. S7-09-22; Release Nos. 33-11038, 34-94382, IC-34529: Cybersecurity Risk Management, Strategy, Governance, and Incident Disclosure. NAM. May 9, 2022. https://www.sec.gov/comments/s7-09-22s70922-20128258290055.pdf

37. NDIA. Re: File Number S7-09-22, Securities and Exchange Commission (SEC) Proposed Rule: Cybersecurity Risk Management, Strategy, Governance, and Incident Disclosure. May 9, 2022. https://www.sec.gov/comments/s7-09-22/s70922-20128385-291288.pdf

In the area of cybersecurity, a corporate board may rely on reporting from an in-house cybersecurity team or external consultants. Relying on non-board member experts should not suggest that a company is unserious about cybersecurity.

If the Proposal is adopted in its current form, the Exchange believes that many companies will prioritize attracting board members with "cybersecurity expertise" in order to demonstrate their commitment to managing cybersecurity risk.

With 7,848 companies filing on domestic forms and 973 FPIs filing on foreign forms during calendar year 2020, the NYSE questions whether there are truly enough individuals with both cybersecurity expertise and other relevant experience to make them suitable candidates for service on a corporate board. If a shortage does exist, the Exchange is also concerned that smaller and medium-sized companies may be disproportionately disadvantaged in attracting these highly sought after individuals for board service.[38]

DDN Comments: The NYSE fails to recognize the board as a part of the system of cybersecurity and they espouse the OUTSIDE EXPERT FALLACY as a suitable proxy for cybersecurity governance — essentially arguing that fiduciary duty can be outsourced. It can't, which is the reason the SEC required resident financial expertise on boards with SOX in 2002. They also use the SHORTAGE FALLACY to argue that small and medium-sized companies would be at a disadvantage in finding cyber experts for boards, so again, why fix it for anyone.

FALLACIES PRESENTED				
ONE SIZE FITS ALL	ONE TRICK PONY	SCARCITY	SLIPPERY SLOPE	OUTSIDE EXPERT
		✔	✔	✔

PPG
PPG is a New York Stock Exchange listed, global manufacturer of paints, coatings and specialty materials with 2021 sales of $16.8 billion.

What They Said: The requirement to disclose whether the issuer has a cybersecurity expert on the Board of Directors could evolve into a market expectation that all issuers have an expert on their Board. PPG does not believe that the Commission's disclosure rules should be a "de facto" governance requirement. Thousands of companies would become subject to this requirement simultaneously, many of which would need to add a new director to their Board. To meet this requirement, issuers may have to create a new seat on the Board solely for a cybersecurity expert. Unlike the experience necessary to be an "audit committee financial expert," the requirements of proposed Item 407(j) are so specific that there likely is not a large pool of director candidates with this level of expertise who also have the general leadership and business experience to serve as a director of a public company. Directors can gain expertise on cybersecurity (or many other company risks) through educational opportunities, table-top exercises and from the issuer's own cybersecurity team. Issuers would be better served having a cybersecurity expert with the qualifications set forth in proposed Item 407(j) on their management team, rather than on the Board.[39]

DDN Comments: Yes, evolving into a market expectation is the point, it would be a collective strengthening of America's cybersecurity resiliency that would strengthen economic growth and output. Relying upon the SHORTAGE FALLACY the PPG CFO also falsely claims that the cybersecurity expertise definitions are too narrow as compared to the financial expertise requirements of SOX. They are very similar actually. He also presents the ONE-TRICK PONY fallacy that director cyber experts would not have general leadership and business experience necessary to be corporate directors. They conclude saying this is exclusively a management responsibility, ignoring the board's fiduciary responsibilities and role as a control in the cybersecurity system.

FALLACIES PRESENTED				
ONE SIZE FITS ALL	ONE TRICK PONY	SCARCITY	SLIPPERY SLOPE	OUTSIDE EXPERT
	✔	✔		✔

R STREET INSTITUTE
The R Street Institute (R Street) is a nonprofit, non-partisan public policy research organization

Footnotes

38. Jarkowski, Hope. RE: Cybersecurity Risk Management, Strategy, Governance, and Incident Disclosure File No: S7-09-22. NYSE. May 9, 2022. https://www.sec.gov/comments/s7-09-22/s70922-20128356-291125.pdf

39. Morales, Vincent. RE: Comments on File No. S7-09-22: Cybersecurity Risk Management, Strategy, Governance, and Incident Disclosure. PPG. May 9, 2022. https://www.sec.gov/comments/s7-09-22/s70922-20128345-291106.pdf

head-quartered in Washington, D.C. Its mission is to engage in policy research and outreach to promote free markets and limited, effective government.

What They Said: It is not clear to us how this information benefits investors or how cybersecurity expertise on the Board is more beneficial than other measures — such as ensuring that cybersecurity risk is thoroughly communicated to the Board; that it is incorporated into strategy; and that decision makers at the highest levels are held accountable for cybersecurity failures.[40]

REAL ESTATE ROUNDTABLE

The Real Estate Roundtable brings together leaders of the nation's top publicly held and privately owned real estate ownership, development, lending, and management firms with leaders of major national real estate trade associations to address key national policy issues impacting real estate and the economy.

What They Said: We have similar concerns about the Proposal's requirement that issuers disclose the cybersecurity expertise of directors and management. We fear that this will pressure companies to hire "cyber experts" of unproven value to the company, simply to "check the box." We note that the Proposal suggests that companies should consider whether directors, or relevant managers, have obtained a certification or degree in cybersecurity, which raises additional concerns. Because there is no commonly accepted credentialing, or credentialing body, for a "cyber expert," the presence or absence of such a designated expert may mislead investors. Further, as many others point out, there is a limited supply of cybersecurity experts of whatever credentials, suggesting that this requirement inevitably sets many companies up for immediate failure.[41]

DDN Comments: Evidence shows that without director cyber experts, boards check-the-box on cybersecurity. The SORTAGE FALLACY is also presented.

SECURITIES INDUSTRY AND FINANCIAL MARKETS ASSOCIATION

SIFMA is the voice of the nation's securities industry. They advocate for effective and efficient capital markets.

What They Said: Requiring disclosure of cybersecurity expertise for a member of the board of directors may be difficult for many registrants to implement—especially depending on the size of the registrant. Board of directors are distinct from management, as the board's role is one of oversight whereas management is required to have subject matter expertise. The Proposal fundamentally changes the role of the board of directors in a way that is untenable. The board should have the flexibility to determine its own composition and take into consideration the collective expertise of the board, holistically. Boards are, by design, deliberative bodies and tasked with oversight of numerous risks — of which cyber is only one of those risks. "Special interest" board members are not the best way to advance the oversight of a registrant or of a board as a deliberative and collective body; therefore, the Proposal could undermine this model. Current disclosures required concerning board's business experience should be sufficient to elicit relevant information for investors. SIFMA recommends the commission provide an alternative method of disclosure, such as whether the board of directors engages in regular cybersecurity education sessions or engages with expert advisers on the topic.

Appointing a cybersecurity expert should be an individual judgment for individual companies in light of their structure, sophistication, and relevant risks. There is also a potential sourcing issue – there may be not be sufficient people with cybersecurity expertise and requisite board skills (in some cases C-Suite level). It is also far from clear that the types of expertise provided as examples (e.g., having a cybersecurity certification or degree) is necessary for effective oversight of cybersecurity matters. This could be misleading and create a false sense of confidence among investors, because a company without board members with cybersecurity expertise may have an extremely knowledgeable CISO and other information security staff. Yet the absence of a board member with specific credentials or past experience may unjustifiably impact the perceived

Footnotes

40. Wechsler, Steven. DeBoer, Jeffrey. Re: File Number S7-09-22: Cybersecurity Risk Management, Strategy, Governance, and Incident Disclosure. Nareit The Roundtable. May 9, 2022. https://www.sec.gov/comments/s7-09-22/s70922-20128355-291124.pdf

41. MacGregor, Melissa. Wagner, Thomas. RE: File No. S7-09-22; RIN 3235-AM89: SEC Proposed Rule: Cybersecurity Risk Management, Strategy, Governance, and Incident Disclosure. SIFMA. May 9, 2022. https://www.sec.gov/comments/s7-09-22/s70922-20128347-291108.pdf

risk exposure of a registrant. The SEC should simply not specify how public companies must go about discharging their cybersecurity responsibilities. We recommend the Commission lighten its language on this requirement to ensure that this does not transform into a mandate for every registrant to have a cybersecurity expert on its board. The Commission proposes requiring disclosure of any cybersecurity expertise of board members and disclosing the names of those directors with expertise. Disclosing the names of board members with cybersecurity expertise could draw unwanted attention to these board members, including from bad actors (e.g., hacktivists) or litigants, and could thereby discourage board memberships.[42]

DDN Comments: Declaring that the proposed rules fundamentally change the board is correct, as it strengthens the board as a control in the cybersecurity system, which is the point. Current approaches have proved inadequate, i.e., self-regulation, which is the reason behind the SEC's proposed rules to begin with. They also claim naming cyber expert directors will put them individually at risk, a situation that exists for existing directors and descriptions of skills included in Proxy statements.

FALLACIES PRESENTED				
ONE SIZE FITS ALL	ONE TRICK PONY	SCARCITY	SLIPPERY SLOPE	OUTSIDE EXPERT
	✔	✔	✔	

VARIOUS INDUSTRY ASSOCIATIONS

A joint letter was submitted from these associations. Several of them also submitted their own letters which are also highlighted in this document.

ACA Connects—America's Communications Association

ACT | The App Association

Agricultural Retailers Association (ARA)

Airlines for America (A4A)

Alliance for Automotive Innovation

American Chemistry Council (ACC)

American Council of Engineering Companies (ACEC)

American Council of Life Insurers (ACLI)

American Fuel and Petrochemical Manufacturers (AFPM)

American Gas Association (AGA)

American Petroleum Institute (API)

American Property Casualty Insurance Association (APCIA)

Biotechnology Innovation Organization (BIO)

Competitive Carriers Association (CCA)

Consumer Technology Association (CTA)

CTIA

Federation of American Hospitals

The Fertilizer Institute (TFI)

Global Business Alliance

Healthcare Information and Management Systems Society (HIMSS)

Information Technology Industry Council (ITI)

Interstate Natural Gas Association of America (INGAA)

National Association of Broadcasters

National Association of Mutual Insurance Companies (NAMIC)

National Association of Chemical Distributors (NACD)

NCTA—The Internet & Television Association

National Electrical Manufacturers Association (NEMA)

NTCA—The Rural Broadband Association

Professional Services Council (PSC)

Securities Industry and Financial Markets Association (SIFMA)

Semiconductor Industry Association (SIA)

Telecommunications Industry Association (TIA)

U.S. Chamber of Commerce

USTelecom—The Broadband Association

What They Said: Company boards should prioritize managing cyber risks but not through SEC mandates requiring cybersecurity "expertise." Our associations advocate for companies to proactively prioritize cyber risk management activities, but they are concerned about the SEC's call for companies to disclose the name of any board member who has cybersecurity expertise. We believe that board experts should not proliferate via government directives. Prescriptive disclosures intended to drive company behavior regarding which subject-matter experts sit on companies' governing bodies could lead to unwieldy and unwanted outcomes (e.g., giving investors a false sense of confidence because of the presence of a board cybersecurity "expert").

Footnotes

42. MacGregor, Melissa. Wagner, Thomas. RE: File No. S7-09-22; RIN 3235-AM89: SEC Proposed Rule: Cybersecurity Risk Management, Strategy, Governance, and Incident Disclosure. SIFMA. May 9, 2022. https://www.sec.gov/comments/s7-09-22/s70922-20128347-291108.pdf

Also, cybersecurity talent is scarce globally. It is unclear where companies would get the cybersecurity experts that would be driven by the Commission's proposed requirement to disclosure such expertise.

There is a well-established lack of cybersecurity talent for the public and private sectors that would impede companies' abilities to recruit board cybersecurity experts. The SEC's proposal could even create unintended barriers for historically underrepresented groups to move into cybersecurity management or board leadership roles—not due to the lack of qualifications but to the absence of formal credentials (e.g., owing to their costs) and other certifications. Even if companies could obtain the relevant cybersecurity experts for board positions, no evidence has been convincingly shown that this requirement would inform investors or improve companies' cybersecurity preparedness.[43]

DDN Comments: This response is copied verbatim from the U.S. Chamber of Commerce response and illustrates the lack of critical thinking and a coordinated propaganda campaign advocating against the SEC proposal for director cybersecurity expertise disclsoure.

IT/CYBERSECURITY COMPANIES

MICROSOFT

Microsoft creates platforms and tools powered by AI to deliver innovative solutions that meet the evolving needs of its customers. The technology company is committed to making AI available broadly and doing so responsibly, with a mission to empower every person and every organization on the planet to achieve more.

What They Said: Although the proposed disclosure creates no explicit obligation to have a "cyber expert" on a company's board, it does create implicit pressure to do so. The rule as proposed invites investors and other stakeholders to question boards with no or minimal cyber expertise among their members as to whether they are equipped to oversee cybersecurity risk.

We are concerned that this trend of requiring disclosure of specific expertise will lead to boards being made up of specialists who will lack an overall picture of the business landscape. The best boards are comprised of highly accomplished generalists with deep intellect and acute critical thinking skills. As a board of directors strives to maintain a diverse set of experiences and attributes, it also expects that each member will be able to understand and contribute meaningfully to the oversight of the broad range of material business, risk, and regulatory issues their role requires them to oversee. Management's responsibility includes educating and communicating to the board in a way that enables effective oversight around these issues. As a result, specific areas of substantive expertise may be beneficial but not required among board members. No board should unduly rely on an "expert;" rather, it should ensure management, or the board's own outside resources, bridge any gaps and answer any questions until the board is satisfied that it, collectively, can effectively discharge its oversight duties.[44]

DDN Comments: Remarkable that Microsoft is on the AGAINST list. They state the obvious that if the board does not have director cyber expertise, investors may question their ability to oversee cybersecurity risk. That is exactly the point of having director cyber expertise on the board, i.e., to strengthen the board as a literal control in the cybersecurity system.

But they believe this is a specialist capability, i.e., the ONE TRICK PONY FALLACY without an ability to see the "overall business landscape picture". They then extend their argument to the OUTSIDE EXPERT FALLACY.

FALLACIES PRESENTED				
ONE SIZE FITS ALL	ONE TRICK PONY	SCARCITY	SLIPPERY SLOPE	OUTSIDE EXPERT
	✔			✔

SAFE SECURITIES INC.

Safe Security is a pioneer in cyber risk quantification.

What They Said: We recommend that it should not be required to disclose the names of persons with cybersecurity expertise on the board of directors.

In addition to the unintended effect mentioned [willingness to serve on a board if named], we believe

Footnotes

43. Various. Re: Cybersecurity Risk Management, Strategy, Governance, and Incident Disclosure (File Number S7-09-22). June 22, 2022. https://www.sec.gov/comments/s7-09-22/s70922-20132693-303184.pdf

44. Scribbr. Fallacies. Retrieved June 28, 2024. https://www.scribbr.com/fallacies/slippery-slope-fallacy/

that the entire board should take joint responsibility for a robust cybersecurity risk management plan, not just an individual board member.[45]

LAW FIRMS

DAVIS POLK

From financial hubs to political capitals to epicenters of innovation, they have more than 1,000 lawyers around the globe.

What They Said: We agree with the Commission that cybersecurity is already among the top priorities of many boards of directors. However, we submit that it does not automatically follow, as the Commission asserts, that investors may therefore find disclosure of whether any board members have cybersecurity expertise to be important. Rather, as with other elements of public company operations and risk management, investors expect the board to exercise oversight of cybersecurity risk management by, among other things, ensuring that management possesses appropriate expertise in the field of cybersecurity, including a suitably management expert where appropriate. However, while the board will exercise oversight of that CISO or other designated management as part of its general oversight of cybersecurity risk management, it is management that applies their expert knowledge day-to-day. The technical background of any specific board member is not inherently relevant to his or her ability to oversee and manage risk, whether that risk is cybersecurity or some other similarly important part of the business.[46]

DDN Comments: Their last sentence is very confusing and illogical and throughout their commentary they are of the opinion that effective oversight is achieved when there is effective management. Their opinion fails to recognize the fundamental role of the board.

HUNTON ANDREWS KURTH

More than twenty years ago, Hunton Andrews Kurth LLP was among the first law firms to establish a dedicated privacy and cybersecurity practice. They regularly work with clients to address a broad range of issues related to cybersecurity and have been called upon to assist with every phase of data breach management, from preparedness and prevention to mitigation and resolution.

What They Said: We urge the Commission to eliminate this disclosure requirement in any Final Rules. Many boards currently lack a dedicated cybersecurity expert, and with good reason—the ability of the board of directors to provide oversight and monitor compliance in a manner consistent with its fiduciary duties is not dependent on having technical subject matter expertise. Boards instead rely on management and outside consultants to develop appropriate cybersecurity policies and procedures, as they are permitted to do under state law. The Commission oversees disclosure by registrants in countless regulated industries, and does not require each board to make disclosure about the technical or regulatory expertise of board members on an industry by industry or discipline by discipline basis, with the critical exception of the audit committee financial expert, which is a requirement derived from a congressional mandate under the Sarbanes-Oxley Act.

With the existing requirement of an audit committee financial expert, alongside the proposed requirement for a board climate expert under a parallel Commission rulemaking, there exists the potential for at least three separate board experts under Commission rules. Given the vastly different skill sets required for these three compliance specialties, we suspect relative few directors would be able to fill all three positions single-handedly. Further, reserving three compliance-oriented board seats would continue to tilt the balance away from strategic-focused boards and more towards compliance focused ones, which is not a trend we believe is in the best interest of investors. While risk management and legal compliance are certainly important functions for any board of directors, we do not believe specific cybersecurity expertise is required to discharge those responsibilities.

Additionally, we believe management and compliance specialists are better suited to tackling mitigating a given risk factor, including cybersecurity.

Footnotes

45. Modi, Saket. Request for comment on its proposed rules, Cybersecurity Risk Management, Strategy, Governance, and Incident Disclosure.SAFE. September 7, 2022. https://www.sec.gov/comments/s7-09-22/s70922-20138346-308377.pdf

46. Davis, Polk & Wardell LLP. Re: Cybersecurity Risk Management, Strategy, Governance, and Incident Disclosure Release Nos. 33-11038; 34-94282 File No. S7-09-22. Davis Polk. May 6, 2022. https://www.sec.gov/comments/s7-09-22/s70922-20128282-290896.pdf

For some companies, cybersecurity risk is not material to the business, and in those circumstances we question the need to have a designated cybersecurity expert on the board.

Naming cybersecurity experts could also have the unintended consequence of placing them personally at risk to malicious action on the part of threat actors. Certain nation states may place named experts under official surveillance or otherwise harass or persecute them without due process. Hacktivists may try to embarrass the named directors by publishing their personal data or taking unauthorized control of personal computing devices in an effort to discredit a given director's cybersecurity acumen. Named directors may also be placed at other virtual or physical risks, and the prospect of such risk counter-productively may serve as a disincentive to board service in the first place.

Moreover, we believe there is a scarcity in the marketplace for technical experts who would satisfy the Commission's proposed criteria for a cyber-security expert. We believe the proposed criteria to demonstrate cybersecurity expertise are unduly narrow, but we fear that even loosening the criteria substantially would do little to alleviate this shortage of human capital. Those that satisfy whatever technical qualifications the Commission settles on may not satisfy the other wholistic criteria that boards consider when recruiting new directors.

The demands on chief information security officers and cybersecurity forensic investigators, two groups who would most likely satisfy any expert criteria the Commission conceives, may leave them with precious little free time to serve on another company's board of directors. Thus, as a practical matter, we believe the population of eligible candidates is too small to ensure that every reporting company has a board cybersecurity expert. Smaller reporting companies and emerging growth companies may struggle to recruit from the limited pool of candidates as those qualified to serve choose the boards of larger or better known companies. While proposed Item 407(j) is crafted so as to not mandate a board cybersecurity expert, we believe most registrants would be hesitant to disclose the absence of one if affirmative disclosure of that fact is required.[47]

DDN Comments: Opening with the statement that fulfilling fiduciary duty is not dependent upon technical subject matter expertise also referring to experts as compliance specialists. For good measure this firm hits almost all of the common fallacies being espoused on this issue and creates a new one. With quite a bit to say on the issues they introduce the FREE TIME FALLACY for CISOs currently working, which is that they would not have time to serve on a board given the demands of their job, thereby contributing to the SCARCITY FALLACY. This observation fails to recognize that over 40% of current S&P 500 CEO's sit on another company's board.

FALLACIES PRESENTED				
ONE SIZE FITS ALL	ONE TRICK PONY	SCARCITY	SLIPPERY SLOPE	OUTSIDE EXPERT
	✔	✔	✔	✔

WILSON SONSINI
For more than 60 years, Wilson Sonsini has represented technology pioneers associated with virtually every milestone innovation, as well as life sciences trailblazers shaping the future of healthcare.

What They Said: We believe that risk should be managed by the board of directors at the enterprise level, rather than prioritizing cybersecurity expertise over other areas of expertise. We agree with the Commission that cybersecurity is a top priority of many boards of directors, as we frequently advise on such risks with board members of our clients. However, board members are expected to manage a wide range of risks that companies face, from cybersecurity risk to geopolitical risks to climate risks to regulatory and litigation risks. Aside from financial expertise, the Commission does not require companies to disclose board members who have specialized expertise in other areas of risk management.

Requiring disclosure of cybersecurity expertise distinguishes cybersecurity as though it is a risk that should be managed differently and not within a broader understanding of enterprise risk management.

Footnotes

47. Hunton Andrews Kurth LLP. Re: Cybersecurity Risk Management, Strategy, Governance, and Incident Disclosure File No. S7-09-22. May 9, 2022. https://www.sec.gov/comments/s7-09-22/s70922-20128371-291226.pdf

This approach goes against established guidance to companies to integrate and assess cybersecurity risks as part of its overall enterprise risk management.

Further, tools, such as National Institute of Standards and Technology's (NIST) Cybersecurity Framework, are designed specifically to enable non-cybersecurity experts to manage cybersecurity risk at the executive level.

We believe that proposed Item 407(j) may cause companies to prioritize cybersecurity expertise on the board of directors to the detriment of other cybersecurity measures and other desired experience. Proposed Item 407(j) also may negatively impact overall cybersecurity across companies. Cybersecurity standards like the NIST Cybersecurity Framework, Center for Internet Security (CIS) Top 20 Critical Security Controls, International Organization for Standardization (ISO) standards, and Service Organization Controls (SOC) all follow a risk-based approach to cybersecurity. The opportunities to enhance cybersecurity programs and purchase more tools and software are limitless while budgets are not. With this disclosure requirement, companies may feel additional pressure to spend more money and resources on having a board member with cybersecurity expertise, even at the expense of other cybersecurity initiatives that might make the company more secure. This requirement may tip the scales on this particular and specific control in a way that does not actually consider the specific risks to the company and the best ways to remediate those risks. A 2022 industry survey recently reported that only 11% of S&P 500 directors disclosed cybersecurity expertise, which suggests there already is a shortage of cybersecurity expertise among board members. This issue may be exacerbated by more companies competing to find members with cybersecurity expertise on their boards of directors, which may make it even more challenging, and thus more expensive, to bring in and retain such board members.

In addition, as companies attempt to create balanced and effective boards of directors with the appropriate experience needed for their specific situation, the required disclosure may put pressure on companies to prioritize candidates with cybersecurity experience over other desired qualifications, whether based on their industry, regulation or shareholder requests.

We believe that Proposed Item 407(j) should be broader, such as providing disclosure of how the board of directors obtains sufficient cybersecurity expertise and expanding on the list of qualifications for cybersecurity expertise.[48]

While we do not contest that the board of directors should dedicate specific time and energy to cybersecurity risk, specific technical expertise on the board of directors is not required to do so. Similar to how the board of directors of a company with broad international exposure may hire a geopolitical advisor to help it assess geopolitical risks, board members may choose to rely on third party experts to answer technical questions, while equally and competently managing cybersecurity risks. As such, we recommend that the Commission consider having companies disclose how boards of directors obtain sufficient cybersecurity expertise, rather than narrowly require a member of the board to have cybersecurity expertise.

DDN Comments: The lawyers in general have a lot to say about why director cyber expertise does not matter along with the belief that cybersecurity risk is a risk like other general risks. This fails to recognize the unique aspects of cyber risk such as its systemic properties, the presence of an active adversary, the dynamic nature of the changing risk landscape, speed with which risk can propagate, etc. They also believe that prioritizing boardroom cyber expertise will draw budget from other initiatives thereby weakening the company — a Fortune 100 corporate director costs approximately $380,000 annually making it one the highest ROI cybersecurity controls a company can implement.

FALLACIES PRESENTED				
ONE SIZE FITS ALL	ONE TRICK PONY	SCARCITY	SLIPPERY SLOPE	OUTSIDE EXPERT
		✔	✔	✔

Footnotes

48. Hunton Andrews Kurth LLP. Re: Cybersecurity Risk Management, Strategy, Governance, and Incident Disclosure File No. S7-09-22. May 9, 2022. https://www.sec.gov/comments/s7-09-22/s70922-20128371-291226.pdf

ACCOUNTING FIRMS

BDO

BDO is an international network of public accounting, tax, consulting and business advisory firms that provide professional services under the name BDO. It is the fifth-largest accounting network in the world.

What They Said: While we recognize the importance of cybersecurity expertise at the board level, we note that boards fulfill the need for 'expertise' in different ways depending on registrant-specific facts and circumstances. For some companies, a board member with requisite cybersecurity expertise may be warranted. For others, while board members understand the registrant's material cybersecurity risks, policies, and incidents, they may supplement their knowledge needs with the use of outside advisors and experts in lieu of having board member 'expertise.' We support the ability of companies, their boards, and shareholders to consider what level of expertise is most beneficial to the organization and how that need will be satisfied (whether internal or supplemental external support for the wide array of topics a board must address). We wonder whether the requirement to explicitly disclose board level cybersecurity expertise may cause companies to seek out members with cybersecurity expertise, even if the board or registrant would benefit more from expertise in other areas.[49]

DDN Comments: Notably, the AICPA is in the FOR camp. And the BDO argument to leave it up to the issuer is being proven to not work before our very eyes, as the vast majority of boards are not taking meaningful steps to strengthen their oversight of this unique, and critical risk. And finally, they present the argument that cybersecurity expertise would create mutual exclusivity with other boardroom needs, which is illogical.

FALLACIES PRESENTED				
ONE SIZE FITS ALL	ONE TRICK PONY	SCARCITY	SLIPPERY SLOPE	OUTSIDE EXPERT
			✔	✔

DELOITTE

The largest professional services organization in the United States with more than 170,000 professionals.

What They Said: We agree with the Commission that board understanding of critical oversight topics is important, including making sure that the CEO hires an experienced C-suite that understands the board's role and how to inform and discuss strategic topics with the board. A board must be informed and knowledgeable enough to advise and challenge management in all areas of its oversight. Boards themselves acknowledged this need, and many use a matrix to define and Identify the skills, experiences, and diversity needed to execute their duties effectively. We have observed that upon identifying needed skills, boards can gain such skills in numerous ways, including by recruiting members with specific experience or by educating existing board members individually or collectively. We note that in addition to proposing required disclosure about whether any member of a company's board has cybersecurity expertise, the Commission recently proposed similar disclosure related to board expertise in climate-related risks. We believe dedicated expertise may be valuable for some companies. In general, however, especially given the limited size of boards, it may not be practical or advisable for a board to recruit dedicated experts in each of its critical oversight areas.[50]

DDN Comments: They are a little bit on the fence here and wishy washy with this as they acknowledge the benefit for some companies but are non-committal to the need for cyber expertise for all companies, which is essentially a leave it up to the issuer position. Exactly what is not working, and why the SEC proposed new rules to begin with. Also, the argument fails to recognize that in a world of distributed risk, a collective security approach is needed. There is a natural incentive for the weak link in a chain to not strengthen itself, but to free ride when fixing their link creates costs for them, but benefits others. I think we can get them to the FOR side.

Footnotes

49. BDO USA LLP. File No. S7-09-22 Cybersecurity Risk Management, Strategy, Governance, and Incident Disclosure. BDO USA LLP. May 9, 2022. https://www.sec.gov/comments/s7-09-22/s70922-20128427-291370.pdf

50. Deloitte & Touche LLP. Re: File Reference No. 57-09-22; Request/or Public Input on Cybersecurity Risk Management, Strategy, Governance, and Incident Disclosure (SEC Release No. 33-11038; the "proposed rule"). Deloitte. May 9, 2022. https://www.sec.gov/comments/s7-09-22/s70922-20133705-303682.pdf

FALLACIES PRESENTED				
ONE SIZE FITS ALL	ONE TRICK PONY	SCARCITY	SLIPPERY SLOPE	OUTSIDE EXPERT
		✔	✔	

CROWE LLP
Crowe LLP is a public accounting, consulting, and technology firm that uses its deep industry expertise to provide audit services to public and private entities. The firm and its affiliates have more than 5,000 personnel in 39 offices.

What They Said: Investor feedback on the relevance of disclosures related to cybersecurity expert representation on boards is important. It is not clear how an investor might interpret an entity's lack of disclosure of a board level cybersecurity expert, but the NACD Cyber-risk Oversight Handbook 202 (Handbook) observes "there simply are not enough 'cyber experts' to populate every board."[51]

DDN Comments: Simply opting out based upon the NACD Handbook (which is written by a third party) and applying the SHORTAGE FALLACY lacks critical thinking and is evidence free.

DEFINITIONS

Definitions for each of the FALLACIES presented are:

ONE SIZE FITS ALL FALLACY – The false claim that the SEC proposed rule defines and dictates one approach for all companies for what constitutes a director with cybersecurity expertise, they do not.

ONE-TRICK PONY FALLACY – The effort to pigeonhole CISO's, or potential director cyber experts, as "specialists" with skills and capabilities that are too narrow to contribute to the broader boardroom and business agenda.

SCARCITY FALLACY – The claim that the generally hyped cybersecurity skills shortage means there also aren't enough qualified cyber experts to go around in the boardroom, so why even bother with disclosure.

SLIPPERY SLOPE FALLACY – The evidence free claim that director cybersecurity expertise disclosure or presence on the board will create negative unintended consequences elsewhere.

OUTSIDE EXPERT FALLACY – The mistaken belief that outside experts or management expertise is a suitable replacement for corporate governance which fails to recognize the board as a control and the simple fact that fiduciary duties cannot be delegated.

Footnotes

51. CROWE LLP. Re: Proposed Rule, "Cybersecurity Risk Management, Strategy, Governance, and Incident Disclosure," File No. S7-09-22. May 9, 2022. https://www.sec.gov/comments/s7-09-22/s70922-20128341-291097.pdf

APPENDIX — DOMINO 24 Thought Leadership

The Importance of Context: Moving Beyond Broad-Based Skills in Director Oversight of Systemic Cybersecurity Risks

Lee Parrish, DDN.QTE
lee@novelsecurity.com

Abstract

In their final rules for cybersecurity, the SEC withdrew the requirement for disclosing cybersecurity expertise at the board level, leaving the board to leverage broad-based skills in risk management without specific cybersecurity expertise to effectively oversee management's efforts. What is missing in the final rules is attention to the importance of *context* in order to carry out even the now diluted aspects of corporate oversight. It is within context, the ability to fully understand the quality of relevance or significance to a particular situation, that effective risk management can take place. If corporate directors do not understand the contextual details of an issue, they are not in an optimal position to question, understand the answers, challenge, and oversee management.

The SEC indirectly supports the importance of context through its repeated use of a relational term: materiality. Exploring the material aspects (details) of an issue lends itself to the idea that those aspects are understood and applied by the decision-makers and those responsible for oversight. We see this notion of applying broad processes to a more specific endeavor in other fields, i.e., philosophy as one example. Instead of using broad themes for making ethical decisions, a more detailed investigation is performed through what is called, applied ethics. Areas such as medical ethics or communicative ethics serve as just two samples of applied ethics. To accomplish a thorough review in an applied manner mandates that all participants understand the details and relevancy of an issue.

Abandoning a broad-based approach to cybersecurity risks, understanding the context will elevate the boardroom discussion and subsequent oversight. While it is important for boards to possess at least one cyber expert on the team, a full board with a general cyber fluency is needed — just as we've seen in the field of finance at the board level.

Submission

Persuaded by public comments to the proposed 2023 cybersecurity rules, the U.S. Securities and Exchange Commission (SEC) agreed in the subsequent final rules that effective risk management of cybersecurity can be accomplished by "directors with broad-based skills in risk management to effectively oversee management's efforts without specific subject matter expertise, as they do with other sophisticated technical matters".[52] While it is true that the core mechanics of effective risk management may remain consistent across varying disciplines, in order to perform truly effective risk management, a detailed understanding of the underpinnings related to a specific risk(s) is required, as well as understanding any specific relief that may be experienced

Footnotes

52. U.S. Securities and Exchange Commission. Cybersecurity Risk Management, Strategy, Governance, and Incident Disclosure, SEC 17 CFR Parts 229, 232, 239, 240, and 249, Final Rule. U.S. Securities and Exchange Commission, July 26, 2023, https://www.sec.gov/files/rules/final/2023/33-11216.pdf.

from the mitigating measures. This is particularly crucial for cybersecurity, in which the risks are systemic and often evolve with much more velocity than other business risks. What lacks in the SEC's view of the effectiveness of a director's broad-based skills in risk management is that addressing the risks at the contextual level produces the most effective risk management and resulting oversight efforts.

Without understanding the contextual details of an issue, to include the likelihood of a risk or how the risk may be realized, (specific to an environment) any risk management efforts and resulting director oversight will, at best, skim the surface of true effectiveness. My view on this topic may lead one to believe that I recommend a transition of risk management duties from management to the board of directors — this is not my intent. Instead, my comments reflect the notion that risk management be a collaborative effort, with both management and the board addressing, or better stated — understanding, the nuances of context as they relate to systemic cybersecurity risks in order to address them more effectively.

We see this notion regarding context in many other fields of study. Consider philosophy, a very broad-based subject that delves into many different ancient disciplines. One such discipline is ethics, where one can find several examples of what is called, applied ethics. Applied ethics focus on varying subjects under the domain of general ethics, but with a more intensive investigation into the context of that specific field; examples could include medical ethics, communicative ethics, and a host of other areas. Many have found that it may no longer be enough to apply the concepts of general ethics across a broad range of situations, instead, the application of ethics to that specific area or context is more valuable. To accomplish this requires that the person exploring applied ethics in a particular field, understands the details of that specific field if the exploration is to be of significant value. Using applied ethics provides focus on realistic contextual situations, without losing the ancient, broad views of foundational ethical frameworks.

The same transition from a broad-based risk management practice to a more applied, contextual risk management process focused on systemic risks in cybersecurity, would increase the value of the director's risk management and oversight efforts. And to accomplish this, like applied ethics, requires the director to possess a more involved understanding of that field, and not a diluted view provided by broad risk management practices.

> The SEC does indeed, albeit indirectly, hint to the importance of context in risk management with another relational term mentioned several times in the final rules: materiality. Similar to context, materiality deals with the quality of relevance or significance to a particular situation. If making a decision involves understanding contextual circumstances, it would stand to reason that managing risks in order to make those decisions, requires that the risks are completely understood.

If an incident is reviewed by the management team and is considered material, it is feasible to conclude that the incident was reviewed extensively and determined that it is of such high contextual significance to the confidentiality, integrity, and/or availability of the registrant's information systems, it should be raised to the board of directors (and for the purposes of the SEC rules, to investors via public disclosure within four business days). Now in the hands of the board so that they can provide oversight of management's actions, the remediation strategy for this extremely critical issue cannot be reviewed, challenged, or guided, using broad-based risk management techniques. Recent breaches across all industry types show tens of millions U.S. dollars of impact to the victim corporations. Consumer trust, operations, market valuation, to name a few, are increasingly impacted as well. Considering the contextual importance of the incident, i.e., materiality, and the potentially severe impact to the registrant from the incident, is a broad-based view of addressing the incident by the board of directors really enough? With so much at stake, it seems appropriate that any oversight or guidance to the management team by the board be done through the lens of fully understanding the context of the incident.

There are views in our industry that promote the idea of boards having at least one cybersecurity expert among their team of directors. Because of the systemic nature of cyber risks, I agree with this view and feel it would elevate the cybersecurity discussion at the highest levels of corporate governance. However, I believe it should not stop there. Of course, a board need not have all of their directors with expertise in cybersecurity, but there would be value in each board member having at least a general understanding of the specific field — enough to be able to question cyber-related strategies, understand the answers provided by management, challenge them, and provide oversight. This is important for (at least) three reasons.

There is value in having at least one cybersecurity expert on the board to be able to investigate more deeply cyber-related risks. However, this proxy of cybersecurity knowledge from the full board of directors to just one director, diminishes the value of collaboration across the full board and the ability to challenge one another on oversight approaches that will be delivered to management. Consider the role of a director with strong expertise in finance, perhaps they served as a Chief Financial Officer for a huge global corporation in a critical industry. The addition of this expert does not remove the responsibility for all board members to understand general finance subject matter. There may be another board member who has devoted their career to manufacturing, but they still understand the fundamentals of finance. They can contribute to board conversations because they understand a balance sheet, identify EBITA influences in the context of the company, and decipher the performance of the company's gross profit margins. All without significant expertise in finance. The same holds true for the critically important discipline of cybersecurity. Retaining an expert is important for deeply specialized topics, but not as a replacement for general cybersecurity fluency across the board.

Systemic cybersecurity risks continually find their way to the top of corporate enterprise risk management (ERM) listings as one of the most significant threats jeopardizing company success. These risks continue to evolve as adversaries adjust tactics and techniques, while at the same time companies engage more in external partnerships and increased complexity resulting from intertwined external networks. Technology is woven into every aspect of business and like the traditional fields of finance and strategy that came long before it, cybersecurity will remain as a critical part of doing business in the future.

A challenge to this notion of more cyber fluency in our boardrooms may be a critic's idea of employing external cybersecurity or audit firms to help translate the details of cyber for the directors. Briefly stepping away from the focus of information security in this paper to a financial one, retaining such firms may not make good economic sense. If reason number two mentioned above is correct, this would presuppose that more third-party engagements are needed due to the constant evolution that is naturally inherent of systemic threats. Each of these engagements come at a cost, and increased frequency of these engagements to keep up with the threats can equate to potentially ongoing high costs for the company. With more cyber fluency (not necessarily expertise) around the boardroom table, these service costs could be drastically reduced (not eliminated).

Context plays an important part in risk management. Without it, risks are identified, analyzed, and managed without fully understanding them as completely as they should be. Contextual details lead to effective action insofar as the clarity provided in understanding them highlights the details necessary to make informed decisions. Perhaps our industry can borrow from our philosophy counterparts and perform "applied risk management" — with the intent to manage risks specifically relevant to cybersecurity? Whatever path is chosen, the idea of context, largely absent from risk discussions at the board level today, must be integrated into our processes for addressing systemic risks if we are to enhance our collective security postures and incident response practices.

Lee Parrish, QTE
lee@novelsecurity.com

How The DDN DiRECTOR™ And RISCX™ Frameworks Align With The NIST Cybersecurity Framework 2.0

Graeme Payne, DDN.QTE
Senior Partner, TBDCyber
graeme@tbdcyber.com

The NIST Cybersecurity Framework ("CSF") is widely used for evaluating and building cybersecurity programs. NIST recently released an updated CSF (v2.0). This paper examines how the DDN's DiRECTOR™ and RISCX™ frameworks can work alongside and support NIST CSF 2.0.

Introduction

In February 2024, the National Institute of Standards and Technology part of the US Department of Commerce, released an update to the widely used Cybersecurity Framework ("CSF"). The original CSF was published in 2014 and since then has been broadly used in government and commercial enterprises. Many corporate boards are familiar with CSF and its use to evaluate the design of a company's cybersecurity program. Since CSF is outcome based it is not dependent on specific technologies and can be easily adapted to any size organization or industry. The recent update known as "V2.0" includes changes to recognize the importance of governance in managing cybersecurity risks.

Since the inception of the Digital Directors Network ("DDN") the organization has focused on elevating the dialog around governance of cyber and digital risk and in particular the need to understand systemic (or contagion) risks. In 2019, DDN published the DiRECTOR™ and RISCX™ frameworks to help corporate directors identify and understand systemic risk issues across complex digital environments.

In this paper, I bring together these two important approaches — one focused on cybersecurity risk management and one focused on systemic digital risk — and show how each complements the other.

About NIST CSF 2.0

The NIST Cybersecurity Framework (CSF) 2.0 is designed to help organizations of all sizes and sectors — including industry, government, academia, and nonprofit — to manage and reduce their cybersecurity risks. It is designed to be useful across organizations of different cybersecurity program maturity and technical sophistication.

The CSF describes desired outcomes that are intended to be understood by a broad audience, including executives, managers, and practitioners, regardless of their cybersecurity expertise. The outcomes are technology-neutral and therefore provide an organization with the flexibility needed to address their unique risks, technologies, and business considerations.

Outcomes are mapped directly to a list of potential security controls for immediate consideration to mitigate cybersecurity risks. Suggestions for how specific outcomes may be achieved are provided in various online resources that complement the CSF.

CSF is organized into six core functions which are then further divided into 22 categories and 110 sub-categories. The Functions organize cybersecurity outcomes at the highest level:

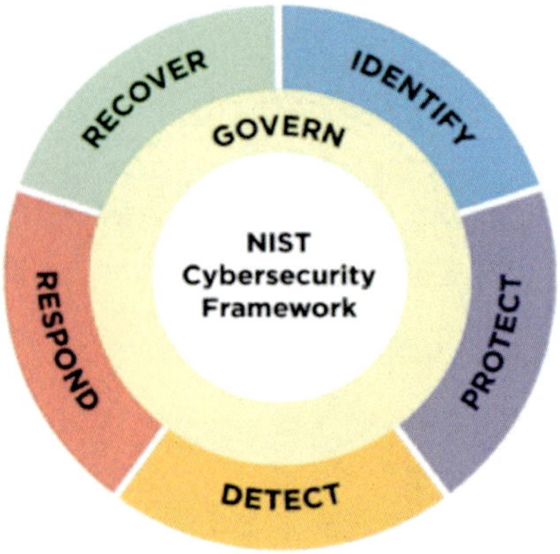

- GOVERN (GV) — The organization's cybersecurity risk management strategy, expectations, and policy are established, communicated, and monitored. Governance activities are critical for incorporating cybersecurity into an organization's broader enterprise risk management (ERM) strategy.
- IDENTIFY (ID) — The organization's current cybersecurity risks are understood. Understanding the organization's assets (e.g., data, hardware, software, systems, facilities, services, people), suppliers, and related cybersecurity risks enables an organization to prioritize its efforts consistent with its risk management strategy.
- PROTECT (PR) — Safeguards to manage the organization's cybersecurity risks are used. Once assets and risks are identified and prioritized, PROTECT supports the ability to secure those assets to prevent or lower the likelihood and impact of adverse cybersecurity events, as well as to increase the likelihood and impact of taking advantage of opportunities.
- DETECT (DE) — Possible cybersecurity attacks and compromises are found and analyzed. DETECT enables the timely discovery and analysis of anomalies, indicators of compromise, and other potentially adverse events that may indicate that cybersecurity attacks and incidents are occurring. This Function supports successful incident response and recovery activities.
- RESPOND (RS) — Actions regarding a detected cybersecurity incident are taken. RESPOND supports the ability to contain the effects of cybersecurity incidents. Outcomes within this Function cover incident management, analysis, mitigation, reporting, and communication.
- RECOVER (RC) — Assets and operations affected by a cybersecurity incident are restored. RECOVER supports the timely restoration of normal operations to reduce the effects of cybersecurity incidents and enable appropriate communication during recovery efforts.

Systemic Digital Risk

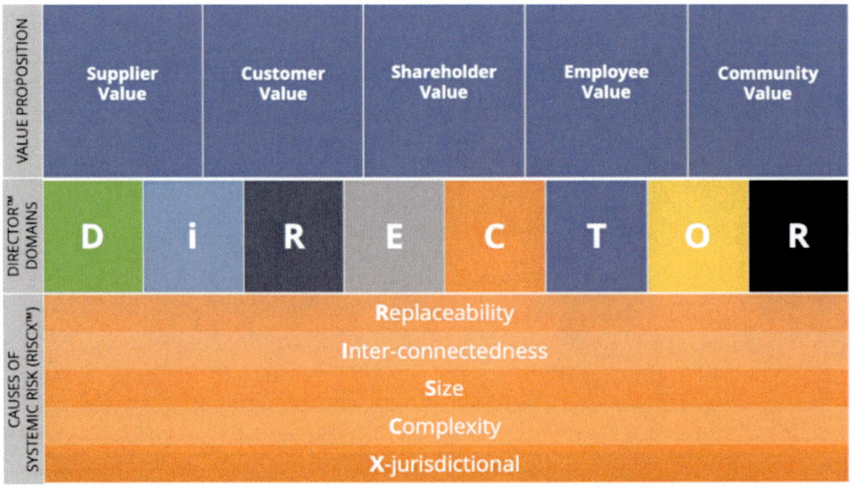

Systemic digital risk is the threat that a component failure or weakness will jeopardize the larger digital and business system through contagion effects. Corporate directors need to understand systemic risk in order to govern it. For many Boards governing cybersecurity and digital risk is a relatively new experience.

With the continued rapid increase in digitizing most business processes and the ubiquitous use of technology and data, corporate directors need frameworks and models to help identify and govern digital risk.

Digital systems give rise to potential adverse impacts (such as cybersecurity incidents) but also can deliver tremendous upside and value to a business.

Corporate directors need to understand risks and potential adverse events as well as how to maximize the value of the investments in digital systems.

To help further the understanding of systemic digital risk, DDN has created two key and inter-related frameworks. Business value sits at the top of these frameworks as this is the primary role of the Board — to preserve and enhance shareholder value. The DiRECTOR™ framework identifies eight core digital domains that work together across complex digital systems to produce and preserve value.

The eight domains are:

- **Data** - the importance of understanding the value of data and the risks associated with capturing, storing, and using data both while it is at rest, and as it moves through the digital and business system.
- **iArchitecture** - the technology design and backbone of the digital system and how it interacts with the business system. Digital systems embody the software, network, people, business processes and hardware components that enable any business system.
- **Risk Communications** - the strategic and tactical communications that are needed to reach every stakeholder internally and externally.
- **Emerging Technologies** - New digital tools are constantly emerging that can enable, threaten or redefine existing value drivers. These tools also layer new complexities across the existing system. They can also exploit or alter fundamental economic concepts and principles in new ways making them an essential concern in digital risk oversight.
- **Cybersecurity** - integrated processes, technology, and people to identify, evaluate, monitor and respond to cybersecurity risks in the digital system.
- **Third-Party** - Systemic risk compounds across an extended business network that is constantly connected and in motion. Every third-party interacting with a digital system creates interdependencies and compound contagion risks across all systems.
- **Operations of IT** - The pace of change in digital and IT is constant. The ability of an organization and its digital operations to effectively adapt to this type of dynamic environment plays a critical role in the sustainability and resiliency of the entire system.
- **Regulation** - Digital regulation is accelerating and expanding. Fines, penalties, and rules imposed by regulators directly influence business and shape corporate governance practice and policy. Every organization also can engage in public-private dialog on digital legislation.

Systemic risk is primarily caused by five key elements that impact the domains. The RISCX™ framework identifies these five elements:

- **Replaceability** - How rare, irreplaceable, and important is a component within the system? Critical, hard to replace components create significant risk and brittleness in a complex system. The more dependent the system is on the component and the harder it is to replace, the greater the level of risk to the larger system.
- **Interconnectedness** - Digital environments can be highly connected both internally and externally. The more connection points across the system, the greater the risk of contagion, both inbound and outbound and the greater the risk that the failure can spread and cascade.
- **Size** - Size impacts total exposure across the system. The larger the data universe that is being enabled, the more risk there can be within that data and across the system. Size factors from multiple perspectives including end-users, third-parties, software systems, etc. Scale creates more points of vulnerability, adds to digital complexity, and contributes to systemic risk.
- **Complexity** - Complexity in any environment makes it more difficult to understand how the environment functions and how it can be controlled. Digital complexity develops from multiple areas including the number of component elements and heterogeneity within the system and operation. Complex digital environments and operations cause systemic risk to increase.

- **X-jurisdictional** - Companies and networks that span the globe create vulnerabilities from anywhere that can travel rapidly to everywhere. The more regulatory environments, whether government or user imposed, that a digital system is exposed to the more difficult it is to oversee and manage regulatory risk.

How CSF and the DiRECTOR™ Framework Work Together to Identify Systemic Digital Value Drivers and Degraders

Recall that the DiRECTOR™ framework is designed to help corporate directors and practitioners identify and understand how complex digital systems can drive or degrade business value. CSF outcomes can be readily mapped to the framework as shown in Table 1.

Let's illustrate with an example. Think about a large online marketplace that connects merchants with consumers. The marketplace company provides end-to-end support for the sale and delivery of consumer products. Now we will apply the first three domains of the DiRECTOR framework to this example.

Data – the online marketplace collects, stores and uses significant amounts of data about consumers and their buying preferences. It also has competitive data and sophisticated analytical models to help in pricing. For product distribution it uses machine learning algorithms to optimize the distribution network. Some risks that arise include:

- Compromise of consumer PII and sensitive data
- Loss of intellectual property (pricing and distribution algorithms)
- Integrity of decisioning data
- Availability of data to ensure 24x7x365 operations.

Understanding the data security controls in place to mitigate these risks would help determine the nature of these risks and the potential impact on business value. CSF Protect: Data Security category would provide an under-standing of data security controls in place.

iArchitecture – the online marketplace utilizes a network of integrated cloud services to provide round the clock access for consumers and merchants. The architecture requires high degrees of elasticity especially during peak periods. The architecture is complex supporting broad functions such as online payments, billing, returns, warehousing, and distribution. The systems must deliver constant analytics and optimization across multiple locations. Potential risks include:

- System uptime and availability
- Scaling up and down for consumer buying patterns
- Integrity of billing and financial transactions
- Ensuring access to systems and services is appropriate
- Ability to quickly detect and respond to incidents

Each of these risks could drive significant degradation of business value.

Many of the CSF categories could assist in further understanding risks and controls. The most common ones would include Protect: Identity Management, Authentication, and Access Control; Protect: Platform Security; Protect: Technology Infrastructure Resilience; and Detect: Continuous Monitoring

Risk Communications – some of the risks in this domain might include:

- Inability to identify and mitigate emerging risks
- Employees not understanding their role in protecting sensitive information
- Management and Board not aware of cyber risk profile
- Risk acceptance beyond the company's risk tolerance

CSF Govern: Risk Management Strategy and Protect: Awareness and Training categories would provide useful information to help understand and further evaluate the risk communications domain.

Table 1 - DIRECTOR™ and CSF Linkage

DiRECTOR™ Domain	NIST CSF 2.0 Function: Category	NIST CSF Desired Category Outcomes
Data	Protect: Data Security	Data are managed consistent with the organization's risk strategy to protect the confidentiality, integrity, and availability of information.
iArchitecture	Protect: Identity Management, Authentication, and Access Control	Access to physical and logical assets is limited to authorized users, services, and hardware and managed commensurate with the assessed risk of unauthorized access.
	Protect: Platform Security	The hardware, software (e.g., firmware, operating systems, applications), and services of physical and virtual platforms are managed consistent with the organization's risk strategy to protect their confidentiality, integrity, and availability.
	Protect: Technology Infrastructure Resilience	Security architectures are managed with the organization's risk strategy to protect asset confidentiality, integrity, and availability, and organizational resilience.
	Detect: Continuous Monitoring	Assets are monitored to find anomalies, indicators of compromise, and other potentially adverse events
Risk Communications	Govern: Risk Management Strategy	The organization's priorities, constraints, risk tolerance and appetite statements, and assumptions are established, communicated, and used to support operational risk decisions.
	Protect: Awareness and Training	The organization's personnel are provided with cybersecurity awareness and training so that they can perform their cybersecurity-related tasks.
Emerging Technologies	Govern: Risk Management Strategy	The organization's priorities, constraints, risk tolerance and appetite statements, and assumptions are established, communicated, and used to support operational risk decisions.
	Identify: Risk Assessment	The cybersecurity risk to the organization, assets, and individuals is understood by the organization.

Cybersecurity	Govern: Roles, Responsibilities and Authorities	Cybersecurity roles, responsibilities, and authorities to foster accountability, performance assessment, and continuous improvement are established and communicated.
	Govern: Oversight	Results of organization-wide cybersecurity risk management activities and performance are used to inform, improve, and adjust the risk management strategy.
	Govern: Policy	Organizational cybersecurity policy is established, communicated, and enforced.
	Identify: Risk Assessment	The cybersecurity risk to the organization, assets, and individuals is understood by the organization.
	Identity: Improvements	Improvements to organizational cybersecurity risk management processes, procedures and activities are identified across all CSF Functions
	Detect: Adverse Event Analysis	Anomalies, indicators of compromise, and other potentially adverse events are analyzed to characterize the events and detect cybersecurity incidents.
	Respond: Incident Management	Responses to detected cybersecurity incidents are managed.
	Respond: Incident Analysis	Investigations are conducted to ensure effective response and support forensics and recovery activities.
	Respond: Incident Mitigation	Activities are performed to prevent expansion of an event and mitigate its effects.
Third-Party	Govern: Cybersecurity Supply Chain Risk Management	Cyber supply chain risk management processes are identified, established, managed, monitored, and improved by organizational stakeholders
Operations of IT	Govern: Roles, Responsibilities and Authorities	Cybersecurity roles, responsibilities, and authorities to foster accountability, performance assessment, and continuous improvement are established and communicated.
	Identify: Asset Management	Assets (e.g., data, hardware, software, systems, facilities, services, people) that enable the organization to achieve business purposes are identified and managed consistent with their relative importance to organizational objectives and the organization's risk strategy.

	Protect: Technology Infrastructure Resilience	Security architectures are managed with the organization's risk strategy to protect asset confidentiality, integrity, and availability, and organizational resilience.
	Recover: Incident Recovery Plan Execution	Restoration activities are performed to ensure operational availability of systems and services affected by cybersecurity incidents
Regulation	Govern: Organizational Context	The circumstances — mission, stakeholder expectations, dependencies, and legal, regulatory, and contractual requirements — surrounding the organization's cybersecurity risk management decisions are understood.
	Respond: Incident Response Reporting and Communication	Response activities are coordinated with internal and external stakeholders as required by laws, regulations, or policies.
	Recover: Incident Recovery Communication	Restoration activities are coordinated with internal and external parties.

Key Actions for Users of DiRECTOR™

- Security leaders and risk professionals can use the framework to better understand how complex systems can drive or degrade business value. Apply each domain of the Framework against the complex digital system and develop potential risk scenarios ("what could go wrong"), including the potential business value impact. Then use CSF outcomes to identify the specific control measures and risk mitigation procedures (people, process and technology) that your organization has in place to reduce the impact. Evaluate the results against your organization's risk appetite and risk tolerance.
- Corporate directors should ask their security and risk managers to articulate the risks in their organization's complex digital systems and how these are currently mitigated or managed. These should be updated periodically as threats, technologies and complexities in the digital system are constantly changing.

How the RISCX™ and CSF Frameworks Work Together to Help Identify Systemic Risk

The RISCX™ model identifies the five key causes of systemic risk within and across the domains of the DiRECTOR™ framework. Table 2 identifies specific sub-categories within CSF that could provide information about the cause and/or the capabilities the organization has in place to identify systemic risks.

Table 2 - RISCX™ and CSF Linkage

RISCX™ Key Cause	Potential Applicability to NIST CSF
Replaceability - How rare, irreplaceable, and important is a component within the system?	GV.SC-04: Suppliers are known and prioritized by criticality. ID.AM-04: Inventories of services provided by suppliers are maintained.

Interconnectedness – How connected is the system both internally and externally?	GV.SC-05: Requirements to address cybersecurity risks in supply chains are established, prioritized, and integrated into contracts and other types of agreements with suppliers and other relevant third parties. ID.AM-03: Representations of the organization's authorized network communication and internal and external network data flows are maintained. DE.CM-06: External service provider activities and services are monitored to find potentially adverse events.
Size – How large is the system and data universe?	GV.SC-07: The risks posed by a supplier, their products and services, and other third parties are understood, recorded, prioritized, assessed, responded to, and monitored over the course of the relationship.
Complexity – How difficult is the system to understand and control?	GV.SC-07: The risks posed by a supplier, their products and services, and other third parties are understood, recorded, prioritized, assessed, responded to, and monitored over the course of the relationship. ID.RA-02: Cyber threat intelligence is received from information sharing forums and sources. ID.RA-09: The authenticity and integrity of hardware and software are assessed prior to acquisition and use. ID.RA-10: Critical suppliers are assessed prior to acquisition.
X-jurisdictional – How many and how complex are the regulatory environments a digital system is exposed to?	OC-03 – Legal, regulatory and contractual requirements regarding cybersecurity – including privacy and civil liberty obligation – are understood and managed.

Key Actions for Users of RISCX™:

- Security leaders should work with the business and IT colleagues to identify, document, and evaluate end-to-end business processes and data flows for systemic risks. Look at major business processes such as the customer value journey or the product development lifecycle and identify single points of failure (including suppliers, cloud services providers, key infrastructure components, physical facilities).
- Corporate directors should ask management to share their analysis of critical suppliers and how they have mitigated the risks of single points of failure in key business processes and digital systems.

Conclusion

Many organizations are utilizing CSF as a framework for evaluating and reporting on their cybersecurity program, including in reports to the Board of Directors. The recent updates to CSF has significantly expanded the role of governance in managing cyber risks. Security leaders should use the opportunity to enhance their dialog with the Board on cyber risk management and in particular around systemic cyber risk. The DDN DiRECTOR™ and RISCX™ frameworks are useful tools to help articulate and communicate around systemic risk. CSF can support and complement the use of these models.

Graeme Payne, DDN.QTE
Senior Partner, TBDCyber
graeme@tbdcyber.com

Applying The SEC Cyber Incident Disclosure Rules: A Post-facto Analysis Of The Equifax 2017 Data Breach Using DDN's BRFO Framework

Graeme Payne, DDN.QTE
Senior Partner, TBDCyber
graeme@tbdcyber.com

The 2017 Equifax Data Breach remains one of the largest data breaches in recent years. This paper examines how Equifax's incident disclosures could have been different if that breach occurred in 2024.

Introduction

In the DDN *Case Study: DiRECTOR and The Equifax Breach*, Graeme Payne and Bob Zukis provided an overview of the 2017 Equifax Data Breach that resulted in the disclosure of data for 147 million consumers. In that Case Study the authors concluded that "All eight of the DiRECTOR™ domains came into play during the Equifax breach. Their problems were self-inflicted and caused by multiple failures across their complex digital system that compounded to create a much larger, and more expensive problem."

Equifax was highly criticized for the delay in reporting on its data breach. The incident was first discovered on July 29th and not disclosed until September 7th. Equifax made a voluntary disclosure of the breach as an Item 8.01 (other events) disclosure in a Form 8-K filing on September 7th. Over the next 20 months the company made updated 8-K disclosures related to the breach. In addition, both 10-Q and 10-K filings included details and financial impacts of the breach and updated commentary on risk factors facing the company.

In October 2011 the SEC issued interpretative guidance on disclosure of cybersecurity risks and cyber incidents. The guidance did not mandate any reporting period but rather focused on reporting of material cyber incidents in financial statements and risk factors, and reliance on existing guidance for reporting material information. This represented the current guidance at the time of the Equifax data breach.

In July 2023, the SEC approved new rules that require public companies to disclose material cybersecurity incidents within four days of determining that they are material.

As the Equifax data breach has been widely reported and investigated it is useful to see how the disclosures might have been different if the breach had occurred today under the new SEC cybersecurity disclosure rules.

Brief Review of the Equifax Data Breach

On July 29th, 2017, Equifax experienced a cybersecurity incident following a criminal attack on a web portal that involved the theft of certain personally identifiable information of U.S., Canadian and U.K. consumers. Criminals exploited a U.S. website application vulnerability to gain unauthorized access to Equifax's network.

The unauthorized access of information occurred from mid-May through July 2017. The information accessed primarily included names, Social Security numbers, birth dates, addresses and, in some instances, driver's license numbers. In addition, credit card numbers for approximately 209,000 U.S. and Canadian consumers, and certain dispute documents with personal identifying information for approximately 182,000 U.S. consumers were accessed. The Company's investigation determined that personal information of approximately 19,000 Canadian consumers was impacted and approximately 860,000 potentially affected U.K. consumers were contacted regarding access to personal information.

Equifax publicly announced the breach on September 7th, 2017, some six weeks after discovery of the breach. The Company provided a website for potentially impacted consumers to visit and determine if they were part of the breach. Impacted consumers were allowed to register for free credit monitoring and identity protection services. The company stock price dropped by one-third (approximately US $18 billion).

Following the public announcement several investigations were launched by regulatory bodies including the SEC. In addition, various Congressional hearings were conducted. Various class action lawsuits were filed against the Company.

Equifax Disclosures Related to the 2017 Data Breach

Equifax was highly criticized for the delay in reporting on its data breach. The incident was first discovered on July 29th and not disclosed until September 7th. Equifax made a voluntary disclosure of the breach as an Item 8.01 (other events) disclosure in a Form 8-K filing on September 7th:

> *On September 7, 2017, the Company issued a press release providing important information regarding a cybersecurity incident involving access to certain consumer information. A copy of the press release is attached hereto as Exhibit 99.1 and is incorporated by reference herein.*

The company made additional 8-K disclosures related to the incident over the next 20 months:

2017 Disclosures

- September 25th notifying retirement of Chairman and CEO Richard Smith (not specifically linked to the breach in the disclosure but widely viewed in the press as being related the breach).
- September 28th notifying the plan to offer a new service for consumers to lock and unlock credit reports without charge for life.
- October 2nd notifying and attaching a press release that the forensics analysis has been completed and providing an update on the incident.
- October 11th an Item 7.01 Regulation FD Disclosure regarding the agreement with UK regulators to inform 693,665 affected UK consumers.
- November 3rd an Item 8.01 disclosure regarding the results of a Special Committee of Board of Directors investigation into stock sales by certain executive around the time of the breach.

On November 9th, the Company issued its first 10-Q since the incident and included various disclosures related to the incident:

- Under Commitments and Contingencies statements including a brief statement on consumers impacted and actions taken, report that US $27.3 million of pretax expenses had been incurred through September 30, estimate of the contingent liability to provide free credit monitoring to consumers (estimated range of US $56 million to US $110 million), note that litigation claims and investigations impacts cannot be determined, expectation of significant legal and other professional services expenses in future periods (amounts undetermined) and increased capital and operating costs, insurance coverage less a US $7.5m deductible which had not been recorded as of end of quarter.
- The company also disclosed in its Management Discussion and Analysis section the expectation of negative revenue impact of the incident through 2018.
- Under Other Information: Legal Proceedings — a description of the various legal and regulatory actions resulting from the breach with a note that it was not possible to estimate the amount of loss or range of possible losses.

- Within Other Information: Risk Factors — additional language added to cover fact the company is assessing the impact of the breach and litigation and investigations on its business and operations, remediation and security and IT enhancements will be costly and may not be effective, the incident has had a negative impact on reputation and the company cannot assure it will not have a long-term effects on relationships with customers, revenue and business.

In subsequent 10-Q and 10-K filings the company continued to report on the impacts of the breach and provide updates on its risk factors. Interestingly for the first time in the 2024 10K the risk factors mentioned "material cybersecurity incident" (emphasis added).

2018 Disclosures

- On March 1st, the company released its annual report filing (10-K) that included significant disclosures regarding the incident.
- On March 1st, an 8-K disclosure provided an update on the incident attaching a press release.
- On May 4th, an 8-K disclosure regarding a statement submitted to multiple Congressional committees regarding the incident and enclosing the submission.
- On June 27th, an 8-K disclosure on consent order agreement with multiple State Attorney Generals
- On July 31st SEC issued a response letter requesting the company to expand their disclosure of the incident to include the origins of the incident and the material events that transpired once the company received notice of the vulnerability.

The company responded on August 14th with their view that **they did not believe the origin of the incident was material to investors** but agreed to add some additional language to future disclosures. The SEC responded and reminded the company and its management that they are responsible for the adequacy of disclosures.

2019 Disclosures

On July 22, an 8-K filing reported on a material definitive agreement to settle consumer class actions against Equifax. The company recorded an accrual of US$690 million exclusive of legal and professional services costs in Q1 2019.

The BRFO Process

The 2023 SEC Cybersecurity Disclosure Rules state that companies should have an informed and deliberative process for determining cyber incident materiality. However, the SEC has not provided any guidance on what this process might look like.

DDN has developed a framework that helps companies determine cyber incident materiality. The BLAST RADIUS-FALLOUT (BRFO) process defines a structured process for determining materiality aligned with the SEC requirements.

BLAST RADIUS focuses on the impact of the incident across the eight domains of the DiRECTOR framework. FALLOUT focuses on the **quantitative and qualitative value impairment and the stakeholders who are impacted by those impacts.**

For a more detailed description of BRFO see *Cybersecurity Incident to Materiality: A Deliberative Process for SEC Disclosure Materiality Determination* (available on the DDN website for members).

Applying BRFO to the Equifax Breach

On July 29th Equifax first discovered that hackers had gained access to a dispute portal. The breach was discovered after the Equifax security team deployed a digital certificate to an SSL decrypter. It would later transpire that the digital certificate had been expired for over 10 months. With newfound visibility into this segment of the network the security team identified suspicious activity and ultimately took down the web portal. At this point in the breach timeline Equifax had not determined the extent of the breach. Applying the BRFO at this time:

BLAST-RADIUS

- **Data** – the website and supporting databases contain data of millions of consumers who have filed disputes but the exact number of impacted consumers is unknown
- **iArchitecture** – the forensic activity to determine whether attackers have moved laterally or been able to access other systems has not been determined.
- **Risk Communications** – the portal delivers services required to support the Fair Credit Reporting Act. Equifax is regulated by the Consumer Financial Protection Board and here may be implications of taking down the dispute web portal for an extended time.
- **Emerging Technology** – the underlying system is being replaced and an extended incident could cause delays in deploying the new system.
- **Cybersecurity** – the incident response is underway and a forensic firm is being engaged to determine the incident impact.
- **Third-Party** – approximately one million consumers interact with the portal per month. Extended outage of the portal will increase demand for call center services and/or frustrate consumers.
- **Operations of IT** – the incident is being treated as highly confidential and only limited IT personnel are aware of the incident. Managing ongoing IT remediation may be challenging.
- **Regulation** - expect increase in complaints by consumers to the CFPB dispute portal and likely regulatory investigation once this is announced.

FALLOUT

At this point (early in the incident analysis), we can estimate some stakeholder and impairment impacts:

Stakeholder	Possible Quantitative and Qualitative Impacts
Consumers	• Unauthorized disclosure of their PII • Future costs to protect them from identity theft • Time/costs incurred in repairing any loss resulting from disclosure of their data • Potential litigation
Financial institutions	• Unauthorized disclosure of data related to consumer relationships and their disputes • Time/costs to repair losses from disclosure of data • Potential litigation

Shareholders	• Potential loss of company market value depending on extent of data breach • Future revenue impairment due to reputational damage • Increased future costs for remediation and restitution. • Costs associated with potential litigation and regulatory investigations • Increased cost of capital
Employees or Directors	• Costs associated with potential litigation and regulatory investigations

I would argue that as of July 31st there are too many unknowns related to the FALLOUT of the breach to determine at this point if it is material under the new SEC disclosure rules. At one end of the range the exfiltrated data could just be a few hundred or thousand consumers. On the other end of the range, it could be millions.

On August 11th Mandiant had been able to complete the initial forensics related to the incident and reported to Equifax management that hackers may have accessed large amounts of consumer PII. At this time if Equifax had applied the BRFO process, they would most likely concluded that the incident was material under current SEC guidelines. Equifax would have needed to make an 8-K disclosure within four days of gaining this knowledge. This would have been thirteen days after the initial identification of the breach, and twenty-seven days earlier than they in fact disclosed the breach.

Conclusion

The BRFO is a useful structured process to help companies determine cyber incident materiality. When a significant cyber incident occurs there is often limited information available. As the incident analysis unfolds the incident team will learn more about the scope of the incident and its potential impacts. This may take some time (e.g., days or even weeks) but at some point, in this process there will be sufficient data to determine the probability that an incident is material is sufficient to warrant disclosure. Looking at the facts in the Equifax 2017 data breach, this was most likely thirteen days after the initial discovery of the breach (and twenty-seven days before the actual disclosure). Also, since 2017 there have been improvements in incident response and forensics that would likely accelerate the analysis.

As we have seen in 8-K disclosures so far under the new rules companies continue to withhold data about the estimated quantified material impacts of incidents. In the Equifax incident this would have also been difficult to estimate. In retrospect the company incurred legal settlements of over US $690 million and made additional investments in technology and security improvements over the subsequent years in excess of US $2 billion. It would have been difficult to estimate these thirteen days after the breach was identified. Additional 8-K (plus 10-Q and 10-K) disclosures would have been needed as details of remediation and litigation costs became reasonably foreseeable.

Companies should establish internal processes to evaluate cyber incident materiality as part of the incident management processes. They should also regularly conduct tests and simulations to exercise this process and make necessary updates to the processes as required. The BRFO framework is a useful model to use in building your organization's approach.

Graeme Payne, DDN.QTE
Senior Partner, TBDCyber
graeme@tbdcyber.com

Innovate, Go Fast, Stay Secure - Pick Three?

Lex Crosett, DDN.QTE
Dr. Stephen Kraemer, DDN.QTE

Abstract

Most companies drive digital transformation efforts (77%), but only one in three (35%) are successful. Experts predict that companies that fail to transform will cease to be competitive within seven years. Security risks of all kinds continue to escalate, and the rate and intensity of cyber attacks are increasing just as these vital projects fail.

At the heart of failure, boards often perceive a seemingly irreconcilable tension between security and innovation — between the need to move quickly to counter competitive pressure versus the need for greater control to combat cyber threats. Ironically, this unsustainable tension is unnecessary and merely a product of archaic thinking.

There is, in fact, an "Easy" button for boards and senior management, and it is hiding in plain sight. Excessive failure rates of digital transformations should be an artifact of the past. CIO, CTO, and CISO functions must come together and complement aspiration and execution, and a modern, synthetic approach already exists.

The playbook is simple: 1) synthesize the culture into one that joins security and innovation without distinction 2) adopt lean, agile methods 3) establish small, customer-facing teams that include security experts 4) use public cloud for rapid experimentation with readily available tools to establish security guardrails, and finally 6) organize to promote single-threaded leaders who can drive rapid and highly secure innovation.

Introduction

Organizations face two existential threats — the need to move much faster to remain competitive while avoiding being fodder for front-page security breach headlines. Only the dangers posed by the global cyber threat landscape [security] eclipse the pace, intensity, and significance of achieving relevance and survivability through digital transformation [innovation]. Yet, the combined influence of these threats on strategic success could not be more critical[53].

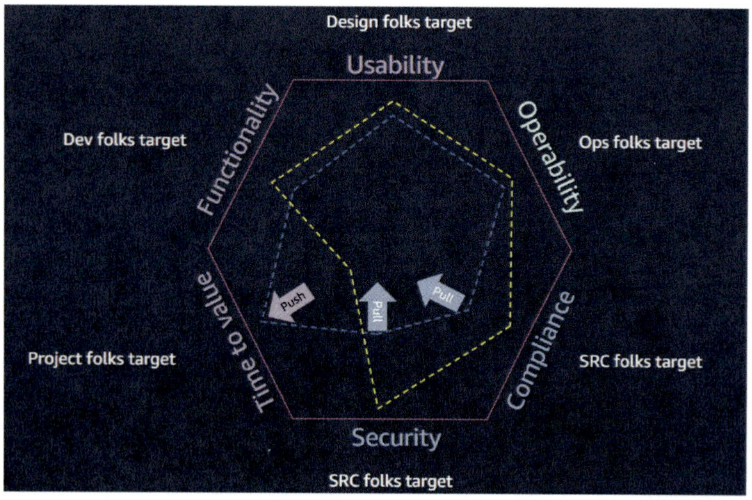

Figure 2 Security v. Innovation Tension, Copyright © 2024, Amazon Web Services

Most directors find the seemingly opposing challenges of innovation [moving quickly] and security [moving slowly and carefully] bewildering and even byzantine, forming complex and mostly adversarial relationships inside the organizations they serve.[54]

A depiction of typical tensions is in Figure 2. There is, in fact, an *Easy* button for boards and

Footnotes

53. https://securityintelligence.com/articles/digital-transformation-balancing-speed-security-innovation-cybersecurity/

54. https://www.nacdonline.org/all-governance/governance-resources/directorship-magazine/online-exclusives/emerging-technology-friend-or-foe/

senior management, and it is hiding in plain sight. A modern, holistic approach to achieve this already exists. In this approach, security and innovation occur together: we will refer to this as Security+Innovation. Excessive failure rates for digital transformations should be an artifact of the past when the CIO, CTO, and CISO functions come together to think organizationally and drive innovative and secure product development.

The playbook for succeeding with Security+Innovation is simple: 1) synthesize the innovation and security culture into one, without distinction 2) adopt lean, agile methods 3) establish small, nimble customer-facing teams that include security experts 4) use public cloud for rapid experimentation and use advanced cloud tools to establish security guardrails, and finally 6) organize to take direction from single-threaded leaders who can drive rapid and highly secure innovation.

Discussion

We posit the notion of an *Easy* button to achieve Security+Innovation. Given the collective human and financial capital globally invested in solving the challenges outlined above, how do we justify the bold claim that this can be made easy?

Cultural Synthesis (and the Easy button)

Research[55] into transformation failure suggests that most transformation efforts fail due to culture[56], squarely placing accountability (and responsibility) with top leadership. Specific details about failure are essential but beyond our discussion here. What's vital is the fundamental Board fiduciary responsibility to influence the creation of a corporate culture capable of securing innovation.

Given that "89% of corporate boards say *digital* is embedded in all business growth strategies, but only 35% of organizations are on track to achieve *digital transformation* goals"[57], boards must carefully influence the methods adopted by management to effect digital transformation as a top-tier imperative. However, board prioritization and control alone don't make the *Easy* button claim a reality, so there is more.

The historical tension between rapid digital innovation and the risk-averse approach commonly employed by many security teams is not without merit. In the 1990s, innovation and profitability were the sole focus of the emerging digital technology experiences powered by the Internet and low-cost, easy-to-consume computing. Security was an afterthought. Only after threats emerged did organizations react, and this reactive pattern meant security was a bolt-on activity; this approach persists and profoundly influences a contentious interplay between security and innovation, a choice between going fast or moving cautiously. In the past, it was pick two (innovate, move fast, stay secure).

However, imagine a reality where embedded and automated security emerges with product releases. In theory, this possibility has always existed but has rarely been pursued. Today, through the evolution of the science of digital technology delivery, there are prescriptive methods for secure, speedy product creation at scale with demonstrable business benefits. We can have our cake and eat it too!

To succeed, organizations must commit to a cultural makeover from top to bottom. The first principle is to establish Security+Innovation as equal partners. This means the organization will simultaneously pursue rapid innovation and robust cybersecurity as a board-level imperative.

Security+Innovation becomes the organization's digital North Star, influencing all investment and strategic

---

Footnotes

55. 7 sins of digital transformation. (2023). CIO. https://www.cio.com/article/656508/7-sins-of-digital-transformation.html
56. https://exdmarketing.smu.edu.sg/acton/media/11523/cultural-transformation-in-the-digital-world
57. 5 surefire ways to derail a digital transformation (without knowing it). (2023). CIO. https://www.cio.com/article/475054/5-surefire-ways-to-derail-a-digital-transformation-without-knowing-it.html

decision-making. With an established culture, 50% of the transformation journey is complete. While the journey's second half may require additional structural change, the industry statistics demonstrate that the alternative to doing so is a fall from competitive strength, stasis, and inevitable decline.

Lean Methods and Organizational Architectures

The fundamental element underpinning the principle of Security+Innovation is organizational architecture. Architecture implies the broadest business sense of meaningful changes to an organization's operating model, including people, processes, and technology.

The decoupling of architecture enables small teams to take ownership of discrete software projects using agile methods while facing and serving customers. This local, decomposed ownership drives accountability and enriches customer interaction. Quality, satisfaction, and security improve as agility plays out across each digital product and service. Additionally, smaller teams innovate more quickly, are unbounded by other teams, and create more secure products through their specialized knowledge and customer-driven diligence.

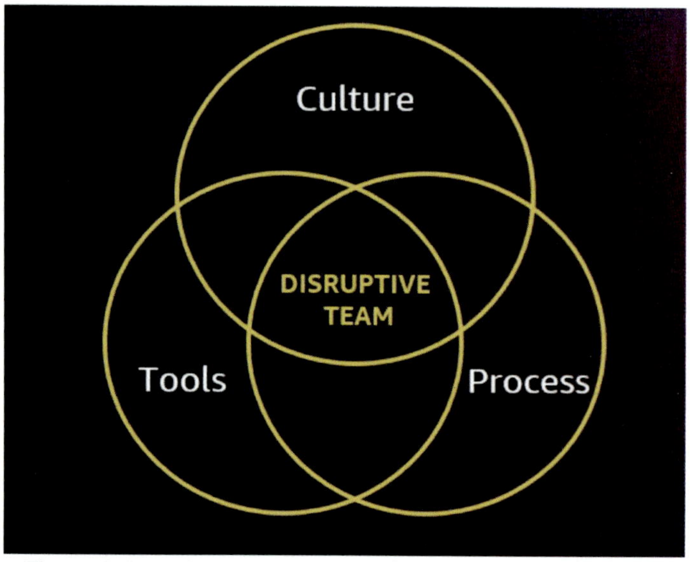

Figure 3 Security+Innovation Architecture and Operating Model, Copyright © 2024, Amazon Web Services Amazon Web Services

Single-Threaded Leaders

To succeed with Security+Innovation, promoting single-threaded leaders to run these teams makes sense. Logically grouped, these leaders maintain ownership of their team's work, and each leader is accountable for business outcomes and can make resource and prioritization decisions.

Armed with the vital tools of synthesis, agility, and cloud computing, this type of leadership combines these elements to establish the organizational basis, direction, and permission to act ongoing: This action is the quaesitum, the *Easy* button.

It is easy not because it is effortless but because it dissolves the tension between security and innovation while simultaneously resolving the organization's digital future. Blended agile teams composed of development, product management, and security professionals led by a single-threaded leader create a position of strength and competitive advantage.

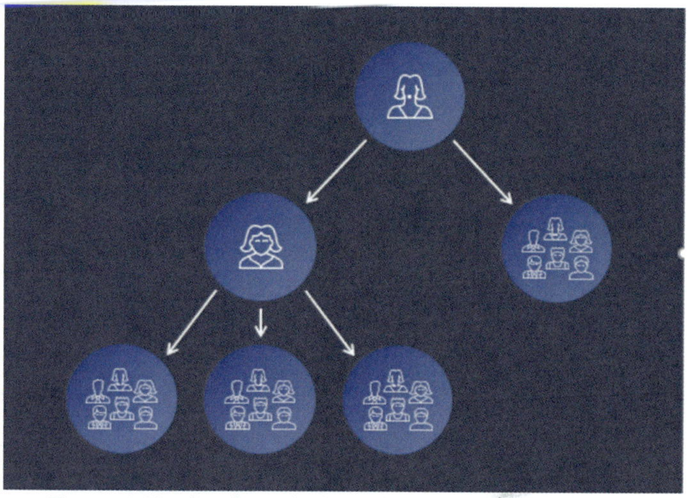

Figure 4 Single-Threaded Leaders, Copyright © 2024, Amazon Web Services

Cloud Computing

Public cloud computing is arguably the most significant development in business transformation, security, and innovation enablement since the inception of the commercial internet. With the cloud, a pay-as-you-go computing model and bundled services allow for low-cost, fail-fast experimentation combined with security and governance excellence.

Using the agile methods and processes discussed previously, organizations can use cloud technologies to accelerate business outcomes, generate deeper insights through data, develop new markets, and outpace their competition - Security+Innovation[58]. For example, companies like GE, Netflix, and Capital One adopted a cloud-first, digital native approach that completely transformed their digital business operations, positioning them as leaders in their verticals.

Research Results

Combining product management, design, development, and security into customer-facing agile teams has been proven compelling for improving delivery speed, security, cost control, and team member retention[6].

DiRECTOR™ Framework

For the DiRECTOR framework, the Security+ Innovation approach spans Operations of IT, Third Party development activities, and Cyber-Security. With a revised set of roles in the organization, the CISO function and the security team are responsible for the many routine activities needed to establish standards, stay secure, and comply with various regulatory requirements. The CISO team can also develop automated, self-service security services and guardrails that each agile team can deploy. The goals of the central and distributed team members remain the same, which are 1) proactively reduce the attack surface of the organization, 2) make security work a seamless part of the development process, and 3) remain ready to change methods, tools, and tactics to stay secure as the threat environment evolves. All of the major cloud platforms make this continuous evolution relatively simple.

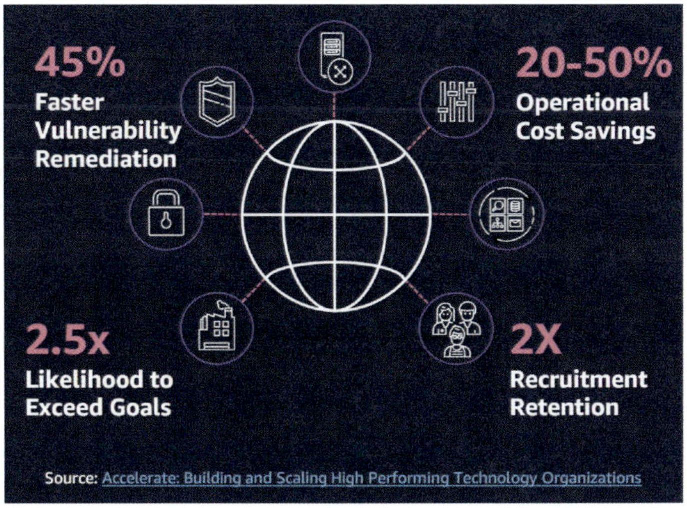

Figure 5 Innovation+Security Benefits, Copyright © 2024, Amazon Web Services

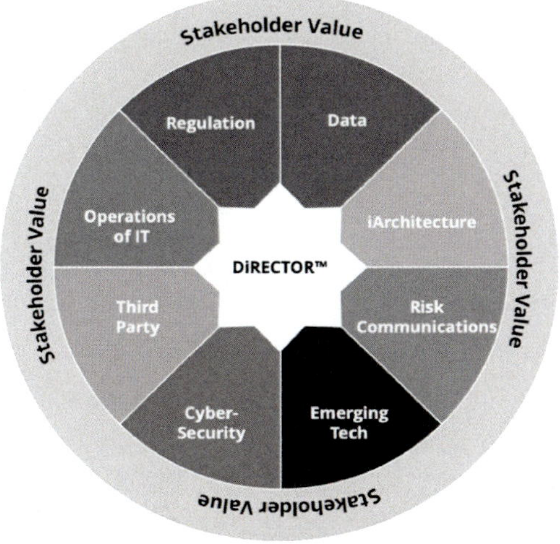

Figure 6 Copyright © 2024, Digital Directors Network (DDN)

The embedded security team members assigned to each agile team should still be responsive to the CISO organization but most intently focused on delivering value to their team's customers.

Conclusion

Some well-established models drive innovation and digital transformation at scale using the Security+Innovation approach. The cost of building new products and services with small agile teams is much lower using cloud technologies, rendering older IT technologies uncompetitive. Because small, customer-facing teams managed with significant autonomy can fail (and succeed) quickly, they can also create customer value more quickly and sustain incremental improvements more rapidly over time.

It is no longer acceptable for Board directors to accept traditional management approaches to product and service development. A forward-thinking board should insist that management become educated on lean, secure innovation methods using cloud technologies and insist on a strategic approach to remaining competitive in this manner.

Footnotes

58. https://www.gartner.com/en/topics/cloud

Additional Reading From DDN

Learn more from these recent publications that detail the SEC's final cybersecurity rules, how leaders harness systemic risk in their vast hyperconnected digital business systems, and the status of governance codes and standards that are emerging around the globe.

The Final SEC Cybersecurity Rules: What's In and What's Out: Impact/Effort Analysis and Comments for Directors and CISOs, by Bob Zukis, founder and CEO, Digital Directors Network, August 2023.

The Great Reboot: Succeeding in a Complex Digital World Under Attack from Systemic Risk, by Bob Zukis, Paul Ferrillo, and Chris Veltos, 2nd edition, 2022. Available on Amazon.

Digital and Cybersecurity Governance Around the World, by Bob Zukis, ***Annals of Corporate Governance, 7:1***, August 30, 2022. Available on Amazon.

DIGITAL
DIRECTORS
NETWORK

DIGITAL DIRECTORS NETWORK

www.digitaldirectors.network

Chicago and Los Angeles
info@digitaldirectors.network

Made in the USA
Las Vegas, NV
27 June 2026

49576813R00057